Astronomer's Observing Guides

Other titles in this series

The Moon and How to Observe It
Peter Grego

Double & Multiple Stars, and How to Observe Them
James Mullaney

Related titles

Field Guide to the Deep Sky Objects
Mike Inglis

Deep Sky Observing
Steven R. Coe

The Deep-Sky Observer's Year
Grant Privett and Paul Parsons

The Practical Astronomer's Deep-Sky Companion
Jess K. Gilmour

Observing the Caldwell Objects
David Ratledge

Choosing and Using a Schmidt-Cassegrain Telescope
Rod Mollise

Mark Allison

Star Clusters and How to Observe Them

With 110 Figures

Springer

Mark Allison FRAS
Mallison@ntlworld.com

British Library Cataloguing in Publication Data
A catalogue record for this book is available from the British Library

Library of Congress Control Number: 2005931470

ISBN - 10: 1-84628-190-3
ISBN - 13: 978-1846-28190-7

Printed on acid-free paper

Printed in the United States of America

9 8 7 6 5 4 3 2 1

Springer Science+Business Media
springeronline.com

In Memory of Trevor for his Words of Wisdom

Preface

Star Clusters contain some of the oldest and youngest stars in our Galaxy. They represent the beginning, the present and the future of the Universe. From the tiniest, sparse open clusters with only a dozen stars, to the giant globular clusters replete with a million stars, observationally, clusters are the most appealing deep sky objects. Through suitable equipment you can witness brilliant blue star forming regions, or densely packed spherical swarms within star clusters, that actually look like the images in those glossy "coffee table" astronomy books. The observer's expectations are therefore matched with the objects appearance in amateur scopes.

Aimed specifically at observational amateur astronomers, this book will provide a comprehensive volume of data, techniques and visual descriptions, with information for all readers regardless of optics used or the individual's experience. In Part I the science of star clusters is uncovered, opening with a general overview of stars, and our galaxy, to set the scene. Star cluster types are then discussed individually, enabling the reader to fully understand each type of object, which adds interest to observing sessions and allows the reader to cross reference this information when tracking down objects in the observing lists. I also hope this text provides some fascinating, and thought provoking reading for the armchair astronomers amongst you, or for those all too often cloudy nights.

Part II describes suitable equipment and accessories for observing clusters, advice and techniques for getting the best out of your kit, and information on planning an observing session. Also included is a comprehensive list of target observations, suitable for all sizes of telescope, and binoculars, to get you started. Sections on recording and imaging clusters, together with catalogues of data, and other resources for planning and research are also included.

This book, like the rest in the series, concentrates solely on the science and observation of star clusters, so you will not find any mythology, history of observing, or space missions within the text. This allows us to get straight into the latest theory on star clusters, which provides the basis for observing these objects later on. It is not a reference book or catalogue, instead it forms a practical observing guide with plenty of specific data to make the observing sessions more meaningful and rewarding.

Even in urban and suburban locations, with light pollution, and dubious weather, many star clusters are much easier to observe and image, than other deep sky objects such as galaxies and nebula, due to their stellar, non-nebulous nature. Previously, as a "general" deep sky observer, I soon realised that I had much more success locating and observing star clusters than any other objects, and my observing sessions quickly turned from frustration into fun. These often neglected objects can be realistically observed and studied, and deserve to be part of any deep sky enthusiasts program. This book will celebrate these attributes, and become your personal guide to the wonders of star clusters.

Amateur astronomers are a noble breed of individuals, who brave the dark and cold nights to embrace the star studded sky, often with the fascination and awe many of us lose in adulthood, but there is a plethora of excellent star clusters "out there" to make it worth your while! Some people may think we amateur astronomers are insane, standing in our backyards in the middle of the night peering at the sky above, so why do we reach for the stars? I believe that in the "faster, better, cheaper" society we live in today, to go out and observe is not only relaxing, but extremely rewarding and awe inspiring. Even the armchair astronomer who does not actually observe can be fully immersed in the amazing world of star clusters, enhancing their knowledge and understanding of the Universe around us, yet he is no less an amateur astronomer.

Did you know that the word "amateur" actually originates from a Latin French phrase, and literally translated means "to love" – what better reason do you need?

Mark Allison

Acknowledgements

First and foremost, I would like to thank Janette Morgan-Allison, my wife and best friend for her support and encouragement throughout this project, without which this book would never have been written.

Sincere thanks also to Maureen Morgan for sparking my initial interest in astronomy with the purchase of my first "real" telescope, and to Jo and Andy Wills for their continued friendship.

I am indebted to Tony O'Sullivan for kindly giving permission to reproduce his entire catalogue of excellent Star Cluster images and observations, and for his help and advice on ideas for the book.

Many thanks likewise, to Cliff Meredith who assisted by supplying many quality cluster images, and for his worthy advice on CCD imaging.

I would also like to acknowledge Daniel Bisque, of Software Bisque for supporting this book with a review copy of TheSky 5 and for permission to reproduce sky charts from the program.

Appreciation is due to Peter Grego who supported my introduction to the publisher and to Darren Bushnall for his advice and guidance on deep sky observing over the years. Both are members of the Society for Popular Astronomy, and deserve credit for the hard work they do on the society's behalf.

Mike Inglis, astronomer, and Series Editor at Springer, helped navigate my initial proposal through the concept and planning stages, which I appreciate wholeheartedly.

Harold Corwin assisted with advice on star clusters, and the NGC/IC Project team in general, especially Steve Gottlieb, generously allowed me to reproduce data and information from their archives.

Thanks also to Andrew James for information on asterisms, and for permission to reproduce data from his "asterism catalogue," and to Phil Harrington, for providing the go-ahead to reference his mini-asterism "discoveries."

Credit is likewise due to Barbara Wilson for allowing use of her material on faint and obscure globular clusters.

I would also like to express gratitude to William Harris, and Wilton Dias for permitting use of their respective cluster catalogues.

Brent Archinal and Brian Skiff also gave invaluable advice on clusters, especially the "nonexistent" variety, and graciously gave clearance to reproduce some of their data.

One of the biggest challenges I faced whilst writing this book was collecting and obtaining authorisation to use material from many sources. I hope I have included everyone that helped me in this quest, and apologise if I have inadvertently left anyone unmentioned. Any errors or omissions are unforgivable, but could perhaps be addressed in a future or revised edition.

As a "mere" observer of these fascinating star clusters, I would like to give full credit to the many professional and amateur astronomers who perform real science, actually make discoveries and work relentlessly to publish and share their catalogues and data. The study of astronomy benefits immensely from this kind of input at all levels, and without it, a book such as this simply could not be produced.

Image Credits

Tony O'Sullivan, deep sky enthusiast and avid CCD imager supplied most of the superb amateur CCD images in this book. They were imaged using a variety of telescopes from a 135mm telescopic lens, to a 4″ f5 refractor, and a 10-inch SCT. These images represent what is realistically achievable from a suburban environment using typical amateur equipment and short exposures between 30 seconds and a few minutes, and have not been subjected to any serious image manipulation.

Cliff Meredith also produced various high quality cluster images for inclusion, using an 8″ SCT telescope and several mono CCD devices.

The professional images are taken from the Digitised Sky Survey (DSS); an all sky digital programme carried out in both hemispheres.

Southern Hemisphere – The use of these images is courtesy of the UK Schmidt telescope, the Particle Physics and Astronomy Research Council of the UK and the Anglo Australian Telescope Board. The Digitised Sky Survey was created by the Space Telescope Science Institute, operated by AURA Inc, for NASA, and is reproduced here with permission from the Royal Observatory, Edinburgh.

Northern Hemisphere – The images from the Palomar Observatory – Space Telescope Science Institute Digital Sky Survey of the northern sky, based on scans of the Second Palomar Sky Survey are copyright of the California Institute of Technology, and are used with their permission.

All finder charts were created in TheSky Version 5 and are the copyright of Software Bisque Inc. www.bisque.com.

Contents

Part I Star Cluster Science

Part II Observing Star Clusters

Part I

Star Cluster Science

Chapter 1

Stars: The Ingredients of Star Clusters

Trying to fully appreciate and comprehend the vastness of Space and the Universe we live in is an almost impossible task, yet we humans have made this one of our most important goals. As we learn more about the Universe, we are also gaining knowledge about ourselves, and our place within it. An example of this is shown in a recent survey, where astronomers released a revised estimate for the total number of stars in the Universe. The results were staggering, with an estimated 70 sextillion stars, that is, 7 with 22 noughts after it, or for the mathematically minded, 10^{22}. It is believed, in fact, that there are more stars in the Universe than there are grains of sand on every beach and desert in the world combined. This figure is, however, only an approximation, based on the number of stars that can actually be detected using the existing telescopes and technology, so this number could be much larger.

Stars are obviously the main component of all star clusters, so it seems appropriate to begin with an understanding of these stellar beacons, and how they relate to, and form star clusters.

Our Star

The Sun, although extremely important to us here on Earth, is "just" an ordinary star, relegated to a mere yellow dwarf or G2 Class. By plotting the Sun's characteristics of color and magnitude on a Hertzsprung-Russell (HR) diagram, the results show that it resides on the Main Sequence, where stars spend the vast majority of their lives. The Sun is, therefore, a middle-aged star, half way through its life cycle. By comparison, stars that reside within star clusters are normally either very old, such as those found in globular clusters, or extremely young, by astronomical terms, like the stars within open clusters, and these can also be plotted on a HR diagram to demonstrate their stellar evolution.

The HR diagram is basically a graph, with star magnitude plotted on the y-axis, and temperature on the x-axis, but this is a powerful tool for astronomers.

Star Birth

Stars are born in stellar factories throughout the Universe by condensing out of clouds of dust and gas, called the Interstellar Medium (ISM). Within this, molecular

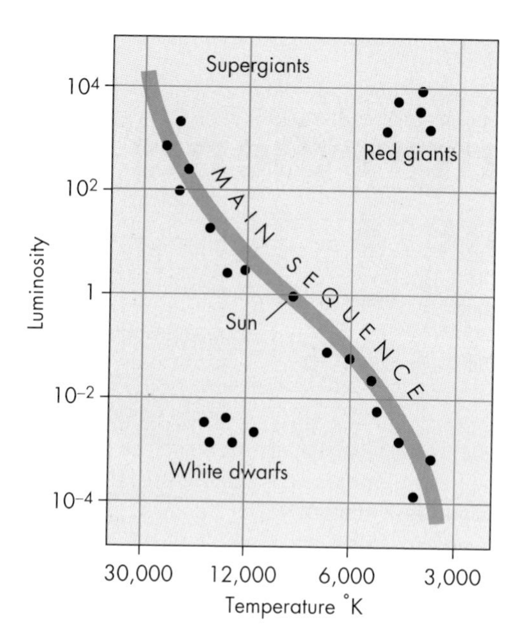

Figure 1.1. A Hertzsprung-Russell diagram showing basic stellar status.

clouds, which are denser than the surrounding matter, begin to collapse under their own gravity. As the temperature and pressure rises, more and more fragments collect and the heat continues rise, until a protostar begins to form. This process, from contracting gas to protostar, takes millions of years due to the vast distances involved. Once the heat and pressure levels reach a critical stage, the protostar begins to emit infrared radiation, and at this point the star is about 10 light-years across. Eventually, the heat reaches over 10 million degrees and the protostar is forced gravitationally into a smaller and smaller space until it becomes typically Sun sized. From here onwards, Helium and Hydrogen are combined, nuclear fusion begins, and the star begins to shine proper.

Stellar Types

Our Galaxy, The Milky Way contains two distinctive star populations, known rather appropriately as Population I and Population II stars. These two types are characterized by their differing ages and chemical content. Population I stars are typically young, hot stars that began life within the last few billion years, and are still forming today. Concentrated within the spiral arms of the galaxy, their orbits are inclined to the plane of the galactic disc. Population II stars are much older, typically 10 to 15 billion years old, so they comprise some of the oldest stars in the galaxy. Located primarily in the central bulge and also in galactic halo that surrounds the galaxy, Population II stars form the basis for the majority of globular clusters, their orbits are erratic and they traverse the galaxy on an elliptical path.

Chemical Content

We have discussed the age of these stars, but what does their chemical content tell us? Using spectroscopy, astronomers can study the stellar spectra radiating from any star to obtain its chemical composition, because each element is absorbed differently.

Spectroscopy is simply the study of the spectra of an astronomical object, to determine its chemical and physical properties. As the spectral lines are radiated from the star, the chemical emissions detected are not sampled directly from the stars' interior, yet there are a suprising number of elements being produced. The main ingredients of all the stars are Hydrogen and Helium, which are combined to produce the vast energy output of light and heat, but carbon, nitrogen, oxygen and iron are all manufactured within the stars. The first elements formed at the beginning of the Universe in the Big Bang were also Hydrogen and Helium, which are described as "light" elements. The other "heavier" elements, all of which are produced from star re-generation, are termed metal elements, and are heavier than Hydrogen or Helium. Astronomers, therefore, refer to the metal content or the "metallicity" of stars, which is high in young Population I stars, and much lower in old, metal poor Population II types. This feature is also demonstrated by the fact that the further a star is from the galaxy, the less metal it contains.

The very first stars should contain no metal, as they formed from the original Hydrogen and Helium in the early Universe, with no regenerated stars to "contaminate" them. But astronomers have found some metal content in all stars, leading to a theorized Population III class, that would make up the very first stars. However, no examples of this type have been detected as yet, which is probably due to their assumed original massive size, which determined a short-life span.

All stars emit radiation at various different wavelengths on the electromagnetic spectrum. These types, starting from the shortest to the longest wavelength are – gamma rays, x-rays, ultraviolet, visible, infrared, microwave and radio waves. By examining stars using these different emissions, astronomers can build up a complete picture of their size, age, temperature and mass. Visual observation forms only a tiny part of the full spectrum, which is why amateur astronomers cannot perform any real science on stars. This also explains why professional astronomers need so many different types of telescopes and space-based observatories to concoct a full picture of stellar objects.

Stellar Magnitude

In terms of brightness, stars are given a number of magnitude expressions, the most prominent for amateur astronomers being the apparent magnitude, which is how bright an object appears from an observer's viewpoint (as opposed to absolute magnitude which uses a standard distance in Space). A star's brightness, however, gives no indication of the distance or size of a star, as we could be looking at a giant star, far away, or a very close dwarf star, that share the same luminosity. Look at the three stars that make up Orion's belt, for example. All three appear very similar in brightness, and they do seem to be connected in the same part of space, but this is simply an optical illusion: a "line of sight" effect. Alnitak and Alnilam, the first and second stars in the belt, are both B type stars with a

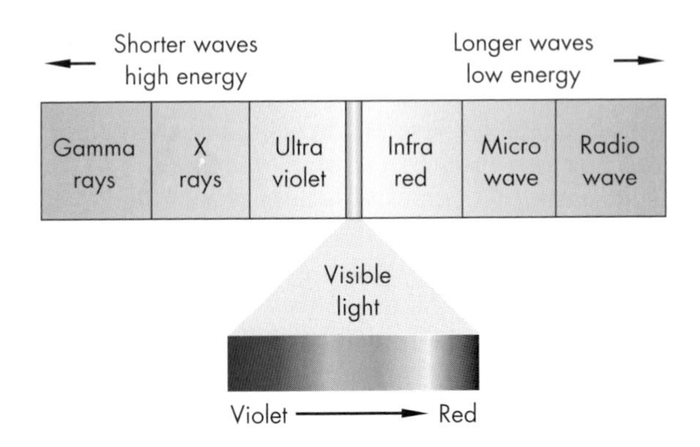

Figure 1.2. The Electromagnetic Spectrum.

magnitude of 1.7, but Alnitak is only 817 light-years away, yet Alnilam is over 1,342 light-years distant. The third star, also a B type, but slightly dimmer at magnitude 2.2, again has a totally different distance of 916 light-years.

Magnitude is a logarithmic measurement based around the number 2.512, where each magnitude is roughly two and a half times brighter than the preceding one, which means that a Magnitude 1 star is 100 times brighter than a Magnitude 6 star. However, this system works negatively, so the brighter an object's appearance, the lower the magnitude. Therefore, objects brighter than Magnitude 1 become 0, or further still, into negative numbers, such as the Sun which is Magnitude –26.

Classification

Color, or the spectra of stars is another important defining factor, that can be applied to most stars, and is directly related to temperature. The hottest stars burn blue and white, whilst cooler stars are yellow, through to orange and finally red. The main spectral classes are:

O – hottest blue star
B – hot blue
A – blue/white
F – white
G – yellow
K – orange
M – red

Type O being the hottest star, decreasing down the scale to M, which are cooler stars. The infamous mnemonic "Oh Be A Fine Girl Kiss Me" is worth repeating here just for its novelty factor, and has certainly helped me to remember the sequence! Astronomers also use subclasses within this structure, by adding the numerals 0–9 after the letter; for instance, the Sun is a G2 star. Other classes have also been introduced to describe more exotic stars, such as Wolf Rayets (W) and cool carbon stars (C), amongst others.

The Luminosity or absolute brightness of some stars is also expressed using roman numerals, which are placed after the spectral type, for example, B7III. The main classes are:

I supergiants
II bright giants
III giants
IBV subgiants
V main sequence dwarves

Within these classes, as can be seen, there are four basic types of stars: main sequence, giants, super giants and dwarf stars.

Main sequence stars are very average in size by stellar standards, and burn Hydrogen and Helium by nuclear fusion within their cores. These stars cover a wide range of colors and temperatures from hot, bright stars, to cooler, dim ones. They are notable for their stability, staying much the same throughout their lives, but once their fuel has expired; their final outcome is largely dependent on their original size.

Giants and Super Giant stars, as the name implies, are massive stars that have begun cooling off, fusing Helium into Carbon in the process. Typically, they are 70 million km in diameter. They are old main sequence stars that have expanded, giving off more light to become giant status. A sun-sized star will become a red giant, eventually blowing off its outer layers to become a planetary nebula, but a red giant finally ends its life as a small faint white dwarf.

Stars that begin their lives much more massive than our Sun (8 times larger or more) will expand even further to become supergiants. These massive stars produce iron in their cores when their fuel expires and their ultimate fate is an incredible explosion, the Supernova, destroying the original star. One possible outcome of the remains is a neutron star, or pulsar, with a supernova remnant cloud, such as the Crab Nebula. Depending on the star's original mass, the final outcome can be even more deadly. A black hole can be produced as the final mass collapses in on itself, so violently that nothing can escape, not even light!

Dwarf stars or low mass stars are main sequence stars nearing the end of their lives. Having burned up most of their fuel, they shine with a feeble light, and are effectively as dead as a star can be!

Distance, Size and Mass

Even the largest of today's telescopes cannot resolve a star into its actual shape. It will always appear as a stellar pinpoint, regardless of the optical aperture or magnification. This is due to the immense distances involved. Even the nearest star to us, Proxima Centauri, is over 4 light-years away, which translates to over 24 trillion miles! To put some perspective on this, if you imagine the Earth and the Sun separated by a distance of one inch, the nearest star would be over 4 miles away. For reference, a light-year is a measurement of distance, not time. It is the amount that light travels in 1 year, which equates to 186,000 miles per second, so 1 light-year is equivalent to 6 trillion miles! However, astronomers would never use miles to measure astronomical distances, the numbers are just too big. In fact, they prefer to use parsecs, where one parsec equals 3.26 light-years.

So, how do astronomers calculate the phenomenal distance, size and mass of stars?

To calculate the distance to stars, various methods can be used, such as parallax, where an objects position is measured over time from two separate points to create a baseline, then the angle of shift can be used to work out the distance mathematically. A simple example of parallax is the apparent shift of an object you see when opening and closing each eye consecutively, when observing a distant object. Due to the extremely small motions of stars, two points on Earth would not show a large enough shift, so astronomers the use the Earth's orbit through Space over a period of one or more years to measure the object's movement against the background stars. For this reason, parallax can only be used for very close samples. More complex methods for calculating stellar distances involve plotting a stars temperature (color) and brightness (magnitude) to find its star type. All stars of a given type shine at known quantities, which are proportional to the distance of the star. Brightness decreases by the distance squared, so astronomers can work out how far away the star must be in order to appear the way it does.

Stars come in all shapes and types, but their size and mass are determined when they originally formed. The largest stars are normally bright but cool such as the Red Giants, whereas smaller stars like white dwarves are hot but typically faint. The heaviest or largest mass stars actually burn their fuel more rapidly, and thus end their lives sooner than smaller mass stars. The Sun appears large here on Earth due to its close proximity to us, but it is a very average star compared to some of the stars throughout the Universe. Stellar sizes are measured by solar radius, where one solar radii (R) is the radius of the sun, which is approximately 1,400,000 km (865,000 miles). Some of the smallest stars, white dwarves, are a hundred times smaller than the Sun, but the largest red giants can be up to a thousand times larger, which is an incredible range of scale. Again, using spectroscopy, a star's size can be deduced by measuring the density of its atmosphere, as different types display significant changes in their spectra. In this way, the spectral type and class of star, such as giant, dwarf or supergiant is found and the size can be determined.

Calculating the mass of a star is much more complex. One solution is based on the observation of binary and multiple stars, which orbit around a common axis. By examining the motions of these stars and their orbits, gravitational equations can be applied which provide evidence for the stars mass. Another method is to use gravitational redshift, which combines measurements of the movement of a star with its radius to determine the results. It has been clearly demonstrated by observation that the luminosity and mass of stars are inextricably linked, where brighter stars from a standard distance are nearly always larger in size.

Star clusters contain many stars of similar type and magnitude, which formed from the same material at the same time, so all their member stars are approximately the same age. Gravity provides the glue that binds theses stellar groupings together. Hot, bright O and B stars are found in their masses in open clusters, but in globular clusters, they are completely lacking. Any blue stars visible in globular clusters are believed to be the product of two older stars fusing together.

When we delve deeper into star cluster types, in the following chapters, these discussions of star types and stellar remnants will become much more relevant and meaningful. As will be revealed, there is much more to these clusters and associations than meets the eye.

Chapter 2

The Milky Way: Home to Star Clusters

Our Galaxy, the Milky Way, is something of a paradox. It is just one of the many billions of galaxies scattered throughout the Universe, a large, but quite normal spiral galaxy in any sense of the word. On the other hand, it is the most important galaxy known, and is unique because it is the only place we know that harbours life.

It is difficult to obtain a clear picture of the Milky Way's structure as we are living on the inside, looking out across one of its spiral arms. We will never be able to travel far way enough to see the Galaxy in its entirety, but to gain some perspective on the size, scale and structure, we can study other galaxies of similar types to fill in the gaps.

By combining these, we have the "best of both worlds," a close-up insider's viewpoint, and billions of distant related galaxies for a "bird's-eye" view comparison.

The Milky Way is like an enormous city, bustling with life, but is also witness to death. Stars are being nurtured to this very day within its nebulous clouds of gas and dust, and these stars are also being dramatically snuffed out, reseeding the galaxy with fresh matter to form new stars. It is a continuos cycle of life, the largest recycling plant you can ever imagine. Even today, 15 billion years after the Galaxy was formed, it is still evolving, constantly changing and regenerating.

Professional interest in our Galaxy and its contents has awakened new interest in recent years as new technology has allowed astronomers to probe and scrutinize the Milky Way as never before. Every tool within the modern astronomer's arsenal has been used to reveal new discoveries in optical, x-ray, infrared and radio wavelengths. However, because the dust and gas within the galaxy obscures our view, there are many regions that are difficult to examine. So, there are probably many star clusters that have eluded us so far.

Galaxy Formation

The general theory for the creation of the Galaxy is in many ways similar to the star formation model explained in Chapter 1, although on a much more massive scale. Dust and gas conglomerated to form a protogalaxy (similar to a protostar) which eventually coalesced to form the globular clusters first in the galactic halo. The spiral arms then formed, and with them, the open clusters, associations and, of course, all the other stars and nebulae. However, there is some controversy as to which parts of the Galaxy actually formed first. The standard theory supports that the halo was the original extent of the galaxy, and that this was created first, from the primordial gas that eventually collapsed in on itself, also demonstrated by the old stars contained within the globular clusters. This matter condensed to create the central bulge, which ultimately began to rotate, creating the disc and spiral arms. This scenario is reasonably confirmed by observations; older stars are found at the outer edges of the galaxy, and young stars towards the galactic center.

New data concerning the placement and distribution of gases in the Galaxy opposes the current models, and suggests that the bulge and the halo formed in a much closer time span, and the disc formed soon after. To confirm this, the halo has also been found to be still accreting new matter from red giant stars that are increasing in mass. It has long been known that our galaxy has fed on stars from other galaxies within our local group (e.g., the Sagittarius Dwarf Galaxy). In fact, one theory goes even further and implies that all globular clusters within the Milky Way are the leftovers of dwarf galaxies that have been cannibalized by our own galaxy, leaving only their cores, which we observe as globular clusters. Nevertheless, a recent spectrographic survey found an alarming similarity in the chemical content of a vast range of halo stars. This challenges the gradual growth by feeding model and implies that most, if not all, halo stars evolved at the same or similar time. To complicate matters further, other studies appear to show that the chemical content of lone and cluster stars within the halo are different to those in dwarf galaxies, further disproving the cannibalization theory. Perhaps a combination of these theories is correct, but this is certainly a hot topic still open for debate.

Size and Scale

It is estimated that there are over 200 billion stars within the Milky Way and its overall diameter is over 100,000 light-years. The Sun lies in the main plane, about two-thirds of the way out from the central hub, at a distance of 26,000 light-years. The Galaxy is in constant rotation, and the Sun takes about 225 million years to complete one orbit, called the cosmic year. Appearing as a band of stars overhead in the sky, from the constellations Cygnus to Vela, the Milky Way is seen in two opposing directions of the local arm, stretching out from the galaxies center. In the Northern Hemisphere, its brighter regions are visible in Cygnus and Aquila, but in the Southern Hemisphere, the areas in Scorpius and Sagittarius are more prominent. From our position in the Galaxy, our viewpoint is across one of the spiral arms towards the galactic center, also in Sagittarius.

Galaxy Components

As we are mainly concerned with star clusters, the most important areas of the Galaxy for our purposes are the spirals arms, situated in the disc, where the open clusters and stellar associations reside, and the outer halo, which provides a home for all galactic globular clusters. The most remote globular clusters are believed to map the original outer extent of the Milky Way soon after its formation, and the Milky Way's shape was originally defined by studying these globular clusters.

The main components of the Galaxy are: the central bulge or nucleus, the disk containing the spiral arms, and the outer halo which envelopes the Milky Way.

The Galaxy is also surrounded by a massive corona of hot gas, the residue of its original formation.

Nucleus and Spiral Arms

Recent research provides evidence for a supermassive black hole, known as Sgr A, lurking within the central bulge. Currently, it is lying dormant, having already snacked on much of the stellar material within the Galaxy, although it may become active again some time in the future. The galactic nucleus or central bulge is 20,000 light-years in diameter and 3,000 light-years thick. The nucleus or core has the greatest proportion of stars in the Milky Way, and these are tightly packed, only 1000 AU apart; so collisions must be quite frequent. A dense star cluster near Sgr A has also been discovered recently.

Astronomers have also detected congregations of stars near the galactic center, and due to the dimensions and size of these carbon stars, a possible barred structure is inferred. This means that our Galaxy may, in fact, be a barred spiral, and not an ordinary spiral, as originally thought. There is, however, much speculation on the actual arrangement and distribution of the stars, and their shape within these arms. Even the size and extension of the spiral structure, and indeed exactly how many arms are involved, is in doubt.

Only recently, a fifth spiral arm was detected by radio astronomers in the outer regions of the disk, but more evidence is required to prove that it is a completely distinct feature, and not an extension of the existing outer arm. Defining the spiral arms from our vantagepoint is difficult, but progress has been made by studying star clusters, and the gas contained within them, as beacons to define their paths. Rather like the "cats' eyes" on a dark road, we cannot see the actual road but we can determine its shape.

The Halo

The outline of the halo is traced by the extent of globular clusters, and they traverse the Galaxy in an elliptical orbit inclined to the galactic disc. Due to the this orbit, halo stars can actually pass right through the disc, but generally spend most of their time far away from the galaxies plane. The halo extends beyond the galactic disk and the globular clusters are arranged randomly within it, above and below the plane of the galaxy. The halo and nucleus are also referred to as the spherical component of the galaxy, as they are aligned in a circular tendency towards the galactic center. Some of the oldest stars in the galaxy are contained in the halo, up to 15 billion years old, but there is very little gas or dust in this region, and hence few new stars are forming. Hot, ionised gas is also predominant in the halo, which emits highly at gamma ray wavelengths.

The Disk

In the halo and central bulge, star formation was high after their initial formation, and has now slowed considerably. In the disk, however, stars are still being formed at a continued, albeit decreased pace. The disc itself is divided into two parts, the thick and thin discs, respectively. In the thin disk many stars had already formed, whilst the disk was still in the process of creation from in-falling gas. Its stellar components were forged from the inner first, to the outer parts last, and the thin disk is probably still forming today. Matter within the thick disc revolves more slowly around the galactic nucleus, and is 3 or 4 times the diameter of the thin disc. The metallicity of the stars in the thick disc is lower than in the thin disc, which means that the stars are much older. Some are believed to be as old as some of the halo stars, at 10 to 12 billion years. Each disc type, therefore, formed at different times and over different time spans. Overall, the thickness of the disk is approximately 1000–2000 light-years in depth. Many hot young stars are contained within the disk, which is why it appears so bright, and a large proportion are assembled in groupings, which typically move in circular orbits. These are the stellar associations with 10 to 100 component stars, and the open clusters containing 100 to 1000 stars or more, but we will discuss these in detail in a later chapter. The vast majority of gas and dust in the interstellar material is contained within the disc, and the disc itself has no discernible definite boundary – its shape and size are different, depending on the wavelength used to view it. As we move further from the galactic center, the concentration of stars also decreases.

Further Studies

As well as the main components of the Galaxy that are optically visible, the Milky Way also contains a galactic magnetic field and many charged particles. There does appear to be a missing element though, as there is not enough observable material in the Galaxy to account for the gravitational forces working on the stars, gas and dust. This missing material, could be the elusive dark matter, believed to account for over 90% of the Universe's content. A halo of dark matter could, in fact, surround the galaxy up to 100,000 parsecs (or 326,000 light-years) from its center.

The Milky Way contains a vast selection of star clusters within its disk and halo that can be observed using amateur equipment. In fact, most of the clusters mentioned in this book reside in our own Galaxy. Of the roughly 5,000 known star clusters, over 1,750 are visible within the Milky Way alone. However, the Milky Way is not unique, and has no exclusivity over these objects. Star clusters appear in all the galaxies that have been observed in any detail, and we will visit these extragalactic clusters in a later chapter.

As with most astrophysical objects, there seem to be many different and conflicting theories on the formation and evolution of the Milky Way and its contents, but this is a healthy situation, proving that real progress is being made. If the current models were not continually challenged, we might still believe that the planets revolved around the Earth!

Chapter 3

Open Clusters

On a clear moonless night you can see a dozen or so open clusters with the naked eye. These stellar groupings are sprinkled against the myriad of background stars and have been known since antiquity. Not until you observe these objects with a telescope or binoculars, however, is their true beauty revealed. Contained within these associations is anywhere from a dozen to several thousand stars of different masses, but with the same age, distance and chemical properties, or stellar type. There are about 1,600 confirmed open clusters, or cluster candidates in the Milky Way alone, ranging from bright, loose, examples with nebulosity like the Pleiades in Taurus, to older and more compact clusters, such as the Wild Duck Cluster (M11) visible in Scutum. This number is only a small sample of the suspected 50 to 100,000 open clusters within the Milky Way, but many of these cannot be detected due to the gas and dust which obscures our view.

Many are too far away from the galactic center and become indistinguishable from the background stars, especially in rich stellar regions. Open clusters are also referred to as galactic clusters, as they appear to hug the vicinity of the disc and the central hub of the galactic equator, but the term open cluster is used more commonly.

What are Open Clusters?

Most open clusters reside within the spiral arms of the Galaxy, and are generally composed of young hot stars. They are loosely associated groupings of physically attracted stars, bound together by gravity. Gas and dust is commonly found within these clusters, and many are surrounded with nebulosity, left over from star birth within the cluster. Even today, new stars are forming and regenerating within the central plane of the disc, though most are middle-aged stars. Open clusters are unevenly distributed and are often found clumped together within and around the central plane. All the stars in an open cluster orbit around one another, and are affected by each other's gravity.

Well-known open clusters include the Beehive Cluster, or Praesepe in Cancer, the Hyades, in Taurus, and the Jewel Box, a southern object situated in the constellation Crux. Open clusters are particularly useful for tracing the extent of the galactic disk, and provide a wealth of information on stellar evolution, as their ages and distances are easier to calculate than single stars. Astronomers can also test stellar models using star cluster data to make assumptions about the past, and predictions about future star formation and evolution. Young clusters are bright and can, therefore, be observed over great distances, and in some rich clusters, star density can be up to a hundred times greater than in the solar vicinity.

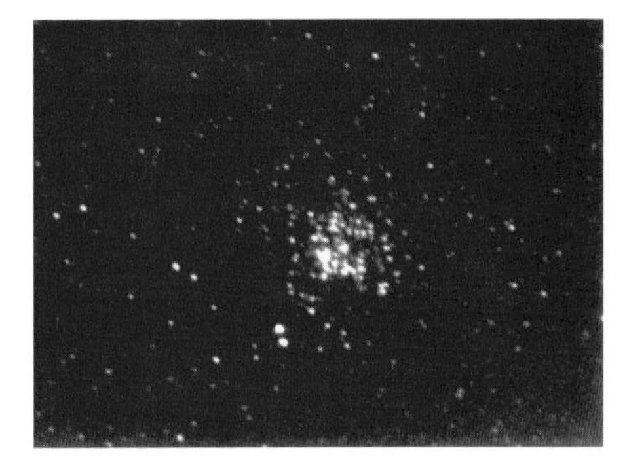

Figure 3.1. M11, The Wild Duck Cluster © Tony O'Sullivan.

How did Stars Form?

The stars within a cluster formed together and are mainly dwarf stars. They were forged from the gas and dust which contracts rapidly under gravity, eventually coalescing into fragments, each fragment eventually forming a star. Young clusters use up most of the dust and gas over time, or this dust and gas can be dispersed by interactions from other massive stars. Clusters form throughout the life of the Galaxy and due to their distribution, they can be used to scrutinize the galaxy on a large scale. A supernova event could also be responsible for the formation of open clusters where they appear in several related locations. Stars within the cluster form in sequence, with low mass stars forming first, and

Figure 3.2. M45, The Pleiades © Tony O'Sullivan.

formation ending with the most massive stars. However, new studies imply that star formation continues even after the high mass stars have been created.

From observation, many open clusters appear to have two distinct regions, the nucleus and the corona, an extended area. The nucleus contains the most massive bright stars, whereas the corona, which is less well understood, is home to the fainter stars. The corona itself appears to define the outer extremity of the cluster, but there is an ongoing debate over whether open clusters started life with two regions, or evolved this way. Some research demonstrates that they originated in this fashion.

Real Clusters?

Unrelated star groupings can easily be mistaken as genuine star clusters as the randomly scattered background stars blend with the clusters, making them difficult to discern. Proper motion, which is the apparent motion of an object, measured by its angular shift in position from the movement of the object through space, can be used on several stars within the cluster to ensure they are true cluster members. Proper motion is essentially a sideways shift that must be measured over a period of several years to obtain any useful data. This is due to the minuscule movements that clusters make over a period of time, but confirms that stars that move together must be related.

Radial velocity measurements, based on the movement of an object relative to the direction of the line of sight, can also be used to determine the distance of cluster members. These measurements are based on the Doppler shifts of the component stars spectral type. The Doppler shift is basically a change in the wavelength of the spectrum where longer wavelengths, called a redshift, indicate an object receding, and shorter wavelengths, known as a blueshift, show an object approaching. Recent proper motion studies reveal that many open clusters contain more stars and are, therefore, larger than their original catalogued data suggests.

Cluster Age

Most open clusters are between 1 million and 10,000 million years old but there are some discrepancies in this threshold. Many clusters are less than 50 million years old. From these figures, an average age of 350 million years is typical. The oldest accurately determined open cluster in the Milky Way is NGC 6791 in Lyra that is estimated to be over 7 billion years old. However, Berkeley 17 in Auriga has often been suggested to be the oldest open cluster, but there is no accurate data to support this claim. New research has suggested that both clusters are very similar in age, and a new figure of 10 billion years has been put forward, so the jury is still out.

Calculating the ages of open clusters is actually easier than for lone stars, so ages and reddening measurements for clusters are more accurate than for single stars. Actual age can be found by observing the brightness or luminosity of the most massive star within the cluster that still resides on the main sequence (i.e., a star that is processing Hydrogen into Helium). High-mass stars burn faster and live shorter lives; so most younger clusters have hot O and B stars within them. Older open clusters contain more red, cool stars and accommodate little metal (elements heavier than Helium and Hydrogen), and this metal content reduces the further away the cluster is from the galactic center.

Cluster age is also determined by examining the radial velocities, mentioned earlier and from the total mass of component stars. Very old open clusters are not commonly found, as their member stars disperse over time, because they are not strong enough gravitationally to stay together for long periods. Old open clusters also contain precious information about the entire life span of the galactic disk and spiral arms to date.

NGC 2362 is the youngest open cluster found so far in the Milky Way. It is aged between 1-2 million years, and its highest mass stars are still on the main sequence.

Cluster Size

Open clusters are composed of far fewer stars than their globular cluster cousins, and are generally much smaller in size. As the edges of an open cluster are ill defined, it is difficult to measure the exact area encompassed, so an average is assumed. This can be in the range from 2 to 75 light-years, but 8–10 light-years in diameter is a rough average. The component stars are fairly tightly packed, with a space of about 1 light-year between them (this actually equates to 6 trillion miles) compared to stars closer to the Sun which are spaced at closer to 7 light-years apart. Looking at the open clusters as a whole, they form a disc about 1,000 light-years thick, and roughly 30,000 light-years in diameter. Many open clusters appear elongated as they begin to age. Stellar forces push on the cluster's interior, forcing it to expand, and as the diameter of the cluster enlarges, the stars become less dense. Clusters of clusters are also found throughout the Galaxy, such as the Double Cluster in Perseus, and groupings in Auriga which include M36, M37 and M38.

Open Cluster Mass

Open clusters often have no determinable shape or structure, and are generally irregular and loosely formed, as opposed to globular clusters, which are spherical in nature and tightly packed with stars.

M18 in Scutum has only a dozen member stars, but M11 is very rich in stars with over 500 components. M67 in Cancer also contains over 500 stars, but these numbers vary dramatically between clusters, and it is almost impossible to give a mean density, although the average range covers between 300 to 4,500 stars. An open clusters mass can be resolved from the amount of bright massive stars contained within it. By comparing these with existing data from close-vicinity stars, the clusters mass is calculated. Typical clusters contain 150 to 200 solar masses, but this figure can be anywhere from 50 to 600 solar masses.

Structure and Composition

Many open clusters contain few faint stars as these low mass components are driven out more easily than more massive stars. Though the majority of stars within an open cluster are normal dwarf stars, their masses can be diverse, and

Figure 3.3. The Double Cluster in Perseus © Tony O'Sullivan.

almost every type of spectral class can exist even though they are technically the same age. Variations in color due to different masses are observed, especially in older clusters, where their member stars have had time to evolve into other stages of stellar life. Several clusters also contain a variety of more unusual stellar types.

For example, "Be" stars are a subdivision of normal B stars, but their spectral emission lines are very strong in Hydrogen, so they are very hot, active stars. "Wolf-Rayet" stars are very rare, unusually hot stars that emit Carbon and Nitrogen formed within a surrounding cloud of gas, due to the stars rapid loss of mass. The surface temperature of these stars is between 20,000 to 50,000 K, but theories on the formation and evolution of these stars are not yet complete.

Shell Stars are another form of B Star and, as the name suggests, they are surrounded by a ring of stellar material, which shows strongly in the stars spectra. This material is possibly created by the rapid rotation of the star, and a typical example is Pleione in the Pleiades cluster. Binary stars are also found at the cen-

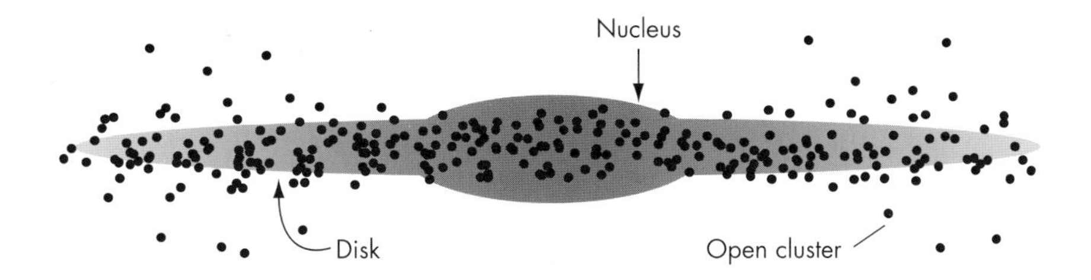

Figure 3.4. Open Cluster Distribution within the Galaxy.

ters of many clusters, and may be responsible for providing renewed energy for other stars within the cluster, and preventing the core of the cluster from collapse.

Neutron stars have also been observed within open clusters, from x-ray observations. These tiny but dense stars, less than 10 kilometres in diameter, have collapsed under their own gravity and are composed almost completely of neutrons. They are much smaller than the Sun but have higher densities, between 1.5 to 3 solar masses. Neutron stars are created from supernova events and can be observed as pulsars using radio observations.

Blue straggler stars have been observed in some of the old open clusters and are a strange breed of old stars that shine with the blue light of young stars, appearring to be unrelated to the cluster. They are, however genuine cluster members, and although their appearance is not fully understood, it is believed they are formed from binary stars that are merging, feeding mass from one another and re-kindling stellar energy.

Origin and Location

As we discussed in previous chapters, the galactic disk is generally assumed to have formed last, so the open clusters contained within it are the youngest members of the Galaxy. Open clusters appear to stretch right across the span of the Milky Way; this is because they do not stray very far from the galactic plane. For our convenience, they are also often grouped together in certain constellations, such as Cassiopeia which contains over 60 candidates of variable size and intensity, and the regions of Monoceros and Puppis that contain many open clusters.

Distance

Cluster distance is computed by comparing apparent (visual) magnitude with the absolute magnitude of known spectral types. However, compensation for the dimming of their luminosity caused by obscuring interstellar gas and dust must be made. Stars are dimmed thus making them appear further away than they actually are. By comparing the actual color observed with the true color or spectral type of the stars, a modification can be made, which is usually in the region of 1 or 2 magnitudes. The most distant old open cluster is Berkeley 29, which has an estimated distance of 22 kpc, or 72,000 light-years, and is thought to be over 3.5 billion years old.

Cepheid variable stars can also be used to calculate the distance of open clusters. These exotic, unstable stars pulsate with a period of between 1 and 50 days. They are extremely bright, yellow supergiants and can, therefore, be observed over vast distances. The relationship between the period and the luminosity of these stars allows their absolute magnitude to be determined. By comparing this with the apparent magnitude, the stars distance can be calculated. Delta cephei was the first candidate of this type to be detected, but many examples of these "standard candles" are now known. A recent study found four cepheid type stars in the open cluster NGC 6910, one of the highest proportions known in the Northern Hemisphere.

Figure 3.5. NGC 6910 (© Tony O'Sullivan).

The closest genuine open cluster in the Milky Way is the Pleiades at a distance of 400 light-years, with the Hyades cluster coming a close second at 1,540 light-years. But we must also consider the Ursa Major moving cluster at a distance of 80 light-years – which is made up of the five inner stars of the "big dipper" in the Plough, although this grouping may be considered more of an association than a true cluster. The term, moving cluster, is used to define a cluster whose member stars are moving in the same direction through Space. The position where these stars converge to a point is parallel to their direction, so distance calculations can be made from this type of study.

A new open cluster, Eta Chamelontis was discovered as recently as 1999. At only 320 light-years away, it is the nearest cluster found within the last 100 years, and was formed about 10 million years ago. It belongs to the cluster group situated in the constellations of Centaurus and Scorpius.

Death of Clusters

Open clusters are short lived, as gravitational forces from within our Galaxy are constantly tugging at their stars, and disrupting the cluster. Encounters with massive binary stars can force out the more remote members of a cluster, and over a period of time they begin to loosen up and drift apart. Eventually, all the member stars within a cluster are dispersed to become part of the general stellar population. Many lone stars were once members of such clusters and our own Sun once belonged to an open cluster that has long since been disbanded. Clusters that reside further out from the central bulge of the Galaxy appear to live longer, so must be less affected by tidal forces from the Galaxy itself.

Star death has also occurred in open clusters. For example, NGC 559, an open cluster in Cassiopeia, is the host of a supernova remnant, the gaseous filaments shedded by a supernova explosion. This cluster contains 60 stars and is over 1,300 million years old, with a distance of 2,900 light-years. It is quite unusual to find supernova remnants within open clusters, as these are the leftovers from a relatively old dying star, yet open clusters are generally quite young.

Classification Systems

The modern open cluster classification, the Trumpler system was devised by R. J. Trumpler, and superseded the older Shapley system. It consists of three main sections:

- Central Concentration – I to IV
  Which denotes decreasing contrast in relation to the star field
  I – detached and strong central concentration
  II – detached and little central concentration
  III – detached and no central concentration
  IV – merges into field, not well defined
- Brightness – 1 to 3
  This is based on the brightness of stars in order of increasing luminosity
  1 – similar or same
  2 – medium range
  3 – vast brightness range
- Richness – p, m, r (poor, moderate and rich).
  The number of stars is portrayed using these letters.
  p – 50 stars or less
  m – 50 to 100 stars
  r – over 100 stars.

A final letter, N is used to denote nebulosity.

For example, NGC 1907 is classed as an II 1 M N cluster.

New Object Discoveries and Developments

One of the main areas of study within open cluster research is regarding star formation, and the cut off point of the main sequence stars. Open clusters are an important source of information because they contain many different star types; so stellar theories can be modelled and tested. There have also been many new open clusters discovered in the past few years, and these clusters have been found to contain some rather unexpected astronomical varieties.

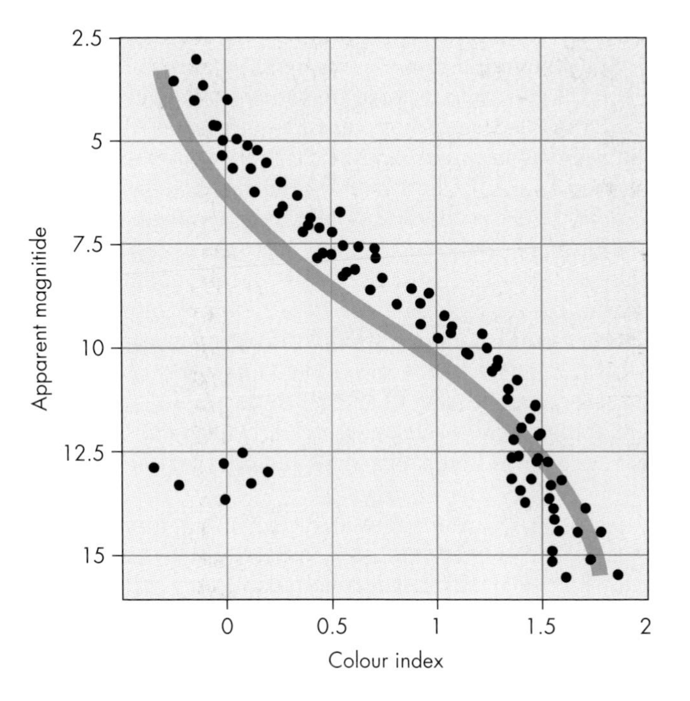

Figure 3.6. A typical Open Cluster HR diagram.

Some open clusters such as NGC 6231 contain eclipsing binaries, which are groups of two or more stars that have varying brightness, caused by the orbit of these stars which allows one to pass in front of the other. This causes periodic dimming and brightening of the star members which can be observed from Earth. A classic example of this scenario, is Algol, the "Demon Star" in Perseus, but this is a single star, not associated with any cluster. M 67 an open cluster in Cancer was newly observed using proper motion studies, and was found to contain a large number of binary stars. Amongst the member stars, the survey detected 23 binaries, of which 4 are spectroscopic binaries, 2 contact binaries and 2 periodic photometric variables. Their x-ray emissions indicate enhanced coronal activity, probably caused by tidally locked rapid rotation of the stars. A blue straggler star with a large eccentric orbit was also discovered, but the survey could not reveal whether its faint blue optical counterpart belongs to M 67. The source could, in fact, be a dormant low-mass X-ray binary, which would be a unique discovery within an open cluster.

A high-precision CCD survey recently undertaken, which observed 20 typical open clusters, revealed over 35 new variable stars whose stellar components vary considerably in mass, age and chemical content. Other CCD-based observing programs conducted using the latest equipment have enabled scientists to more accurately define cluster distance and positions within the Milky Way. For example, new data suggests that NGC 2580 is located at a distance of about 4 kpc and its age is close to 160 million years, and NGC 2588, it is about 5 kpc from the Sun and is 450 million years old. This places NGC 2588 in the extension of the Perseus arm, but NGC 2580 is closer to the local arm.

In 2003, scientists published the discovery of 11 new open cluster candidates that were previously unpublished, all within 500 parsecs of our Galaxy, putting them within the solar vicinity. Several previously catalogued but poorly known clusters were also studied, including Ruprecht 145, and Trumpler 3. The known clusters previously only had approximate coordinates and distances. For the new objects, astronomers initially used stellar density data to pinpoint possible clusters, and then used radial velocity and proper motion information to confirm their cluster status. Finally, photometric sampling was used to determine the color magnitude of individual stars to ensure they were cluster members by defining their type and distance. To test their data analysis, the team compared their results to those from five well-studied open clusters, including Melotte 22, and found them to correlate well with the new discoveries. The 11 new clusters, named Alessi, after their main discoverer are known as Alessi 01, 02, 03, 05, 06 , 08, 09, 10 11, 12, and 19 and are abbreviated in catalogues as Al01, Al02, etc. The discoverers were able to quantify proper motions, distances and ages for the majority of the new candidates.

A recently discovered cluster called Bergeron 1 was found by J. Bergeron in 1997, a tiny sparse cluster, or knot of stars, in Cassiopeia. This was, in fact, a rediscovery, as T. Reiland originally discovered the object in 1988, so it is also known as Reiland 1. The cluster contains only 5 or 6 stars and is not particularly impressive, but proves the point that new cluster discoveries can still be made, even by amateur astronomers.

Recent infrared surveys have allowed astronomers to search for and study clusters within the disk of the Milky Way that is normally obscured by dust and gas. At infrared wavelengths, this dust can be penetrated, and this has led to four new discoveries, called CC11, 12, 13, and 14 (CC stands for Cluster Candidate). CC11 is a compact group of red stars, thought to be a dust-rich young cluster. CC12 is also compact, containing a few bright dominant stars. CC13 lies in the outskirts of the Cygnus Complex, making it a likely young cluster and CC14 was discovered in same field as CC11 as a round grouping of red stars. Surveys such as these help unravel star formation problems, and to answer questions about the spiral structure of the Milky Way, and the age gradients of its member stars. A number of other infrared searches have already been undertaken and the most recent catalogues contain over 270 infrared clusters and stellar groups. Many of them are embedded within dark clouds, such as Tauris-Auriga and Chameleon I. A high proportion of these clusters are double or triple systems, that may prove important in stellar evolution theories. Visual inspection of regions with infrared and radio emissions, and density variation studies can be combined in the detection of new objects.

New clusters observed optically have been found in Scutum, near the galactic center, and have been designated Cluster 1 and Cluster 2 (C1 and C2). They have ages of 25 million years and 500 million years, and their distances have been tallied as 1.64 kpc and 2.2 kpc, respectively. A third cluster found by the same search, called, you guessed it, Cluster 3 was also detected in Canis Major, lying in the direction of the galactic anticenter (the opposite direction to the center!) This cluster is fainter than the other two and is aged somewhere between 31 and 100 million years, with a much larger distance of 4 kpc. Further new cluster discoveries using photoelectric and CCD survey data in the field of the faint Carina Wolf-Rayet (WR) stars WR 38 and WR 38a were recently uncovered. Both WR stars appear to belong to an associated compact cluster of at least six faint OB stars lying at a distance of approximately 14.5 kpc.

Low mass, faint, brown dwarf stars are being uncovered in several open clusters, including the Pleiades, NGC 2516 and Blanco 1. The mass of these small, dim stars has been calculated to be in the region of 30 Jupiter masses, or 0.03 solar masses. A brown dwarf is a cool star that has not accumulated enough mass for nuclear reactions to occur, so it cannot become a true star. Some astronomers believe that brown dwarves are on the borderline between stars and planets; too small to be a star, but too large to be a planet. Several candidates have been studied that contain elements such as water and methane that could not exist in a "genuine" star.

By studying the stellar properties of clusters with known ages and chemical composition, the measurement of rotational velocities of cluster stars is now possible. Multi-object spectroscopy can be used to study the time evolution of the rotational distribution of massive stars. This provides information on the angular momentum in newborn clusters, and the rotational slow down due to magnetic fields or binary companions, and the proportion of increased rotation due to the transfer of mass in close binary stars.

Other unusual information has come to light regarding open clusters in recent years. In 2004, astronomers used the old open cluster NGC 6791 to search for transits caused by possible giant planets occulting stars, betraying their presence. Photometry techniques were used to search for multiple transits, to determine the periods of any long-period variables, and detect any eclipsing binaries. Photometry is the accurate study of stellar magnitudes, calculated using a photoelectric device to examine an object with specific wavelengths. There are several different types of photometric measurement, such as UBV photometry, which measure the ultraviolet, blue and visible wavelengths of stars and groupings. The program observed 20 known variables, and 22 new variable stars were discovered. This search for transit-like events turned up a few single-transit candidates, but further observations would be required to see if these are real planetary transits, or some other phenomenon. The SETI institute also use a catalogue of objects, the "Catalogue of Nearby Habitable Stellar Systems" in the search for possible life. This data set includes 14 old open clusters that will be monitored in their microwave search for technological signals. These clusters were chosen for their closeness to our position in the Galaxy, and their similarity to our Sun.

The most compact cluster of stars in the Milky Way we currently understand, is the Arches cluster in Sagittarius, which contains over 150 young hot stars squeezed into a diameter of only 1 light-year. Several of the member stars are over 20 times larger than our Sun, burning fierce, but short lives, lasting only a few million years. An envelope of hot gas, caused by stellar wind from numerous star collisions has been observed within this cluster that is situated about 25,000 light-years from the Earth. X-ray, infrared and radio wavelengths have been used to probe this cluster, and have shown filamentary structure within the gas that is estimated to be over 60 million degrees in temperature.

The Quintuplet cluster also in Sagittarius is another massive, though less compact open cluster that is similar to the Arches cluster as it also resides close to the galactic centre. It contains the Pistol Star, one of the brightest stars in the entire Milky Way, and the cluster is named after the five prominent stars located in its central region. Both of these clusters have similarities to 'super star clusters', which are covered in chapter 7.

Summary

As with many areas of astronomy that involve formation and evolution theory, open cluster studies seem to be awash with often contradicting information. Many of the distances, ages, sizes and other measurements are estimations, or are based on observations using different wavelengths or varying methods, which can lead to conflicting results. As time goes by, and further studies are made, many of these measurements will be improved and new theories will follow suit, providing us with an even better understanding of these stellar jewels in the night sky. There seems to be no end to the diversity and number of studies being made of open clusters at the present time, and who knows what the next batch of studies will reveal.

Chapter 4

Globular Clusters

Of all the deep-sky phenomena, globular clusters must be one of the most striking objects visible to amateur astronomers. Within their vast spherical orbs lie tens of thousands of ancient stars, tightly packed like a scintillating mass of insects. In some astro photographs, their nucleus appears so dense, you could be fooled into thinking it was almost solid, but with today's telescopes and digital imaging, any cluster can be resolved down to the component stars of its very core. Remarkably, there are just over 150 confirmed globular clusters in our galaxy, which seems an incredibly small number compared to the size of the galactic halo, and the number of open clusters, but there are valid reasons for this, as we shall see later.

The age of globular clusters caused quite a stir in professional astronomy circles only a few years ago, when several prominent examples were dated at 15 billion years old, yet the very Universe they reside in was calculated to be younger, at 13 billion years. Obviously, these numbers did not add up; a daughter cannot be older than her mother. Since then, these figures have been revised because globulars were found to be further away than earlier estimates suggested. If they are more distant, then they must be brighter than originally thought, so they would evolve more quickly, burn faster and, ultimately, be younger. A typical revised estimate for these clusters is now believed to be in the region of 11 billion years, making them younger than their creator, but there are still several stubborn old clusters that refuse to conform to the current theory.

What are Globular Clusters?

Globular clusters define the outer area of the Milky Way; they trace out its original extent from when the Galaxy first formed. They are extremely old, generally massive in size and star numbers, and are distributed fairly uniformly in a spherical inclination around our Galaxy. Stellar numbers are as diverse as the clusters themselves, and can range from tens, to hundreds of thousands of stars, with 100,000 to a million stars being typical. All globular clusters live within the halo, which is the least understood part of the Milky Way, and follow an elliptical orbit around the galactic center. Globulars are fairly symmetrical systems and generally tightly packed towards their core or nucleus, gradually fading out to smaller star numbers in their outer regions. They are almost always linear in nature and spherical in shape.

In the Northern Hemisphere, M13 in Hercules is the brightest, most accessible object, and in the Southern Hemisphere, the clusters Alpha Centauri and 47 Tucanae are two of the brightest and most worthwhile objects. It has been said that once you have seen one globular cluster, you have seem them all, but this a gross

Figure 4.1. 47 Tucanae © Digitised Sky Survey.

exaggeration. Many globular clusters do initially appear similar, but in reality their size and brightness vary dramatically, by a factor of almost a hundred, from the smallest to the largest. The oldest globulars populate the halo, but younger clusters are found closer to the galaxies disc.

How did They Form?

Globular clusters were created from the primary nebulous gas soon after the formation of the Milky Way. As these gas clouds compress, their mass becomes denser and begins to cool down. Eventually, it begins to collect into massive clouds, called giant molecular clouds, and gravity compresses the material further still. After about 100,000 years, the first massive stars are born, pressure and radiation then disperse from the cloud, leaving the newly formed globular cluster behind. Once the cluster reaches 4 million years, a supernova will be created from the most massive stars, and by the 10 million year mark, stellar material will have been shedded as massive stars self destruct. Not until about 20 million years does the globular cluster begin to calm down, to live a more dormant less violent life, typically as we observe them today.

Figure 4.2. Omega Centauri, the brightest globular cluster © Digitised Sky Survey.

It is thought that many thousand of clusters were born at this early stage of the galaxies evolution, though most of these have been dispersed, scattered into the background stars of the halo. Several possible scenarios for their demise have been put forward, which include tidal forces from other stars, and perhaps even collisions with other clusters or galaxies. It has also been suggested that globular clusters may be the remnants of the cores of dwarf galaxies, that have drifted or been pulled into our galaxy. Galactic cannibalization also appears to occur elsewhere in the Universe, and I discussed this scenario within the Milky Way in Chapter 2. Some globular clusters are also slowly rotating, but the mechanism for this motion is not fully understood at present.

One percent of all halo stars are contained with the globular clusters, and it appears that globulars can form wherever there is a particular abundance of gas available for star formation. This figure is in the region of about 25% interstellar matter. Globular star formation is not happening much in Milky Way today, it is a phase of a bygone era, but future collisions with other galaxies from our local group may rejuvenate globular creation over the next few billion years. However, globulars can still form today even though this was previously thought unlikely, as all globulars seem to be ancient. One event that changed this view was the discovery that the globular M54 was the core of the recently discovered Sagittarius dwarf galaxy that is in the process of being cannibalized by the Milky Way. The dual Antenna galaxies also show new globulars forming from their mergers, and these new globular candidates can be as little as 5 billion years old and most result from galaxy collisions and mergers such as this.

Globular Evolution

All globulars are losing stars to the outer halo; they are slowly being ejected from the cluster from perturbations between member stars. These forces can slow or speed the cluster stars' motions, but the gravitational forces holding the cluster together are too strong for it to be completely dispersed.

Although globular clusters are relatively stable objects, they can suffer from collapsed cores. This occurs when massive stars are gravitationally flung to the center of the cluster, where they rebound, only to be thrown back again, like a game of interstellar Ping-Pong! These stars, which are often binary systems, are important as they help prevent core collapse, by creating velocity within the clusters nucleus. As many as 10–40% of all cluster stars may be binaries, and many could have originally formed in this way.

High-resolution studies of the core of NGC 6397 in Ara show an abundance of white dwarf stars that were formed from encounters between clusters stars. NGC 6397 lies at a distance of 8,200 light-years. Measurements over the last century have been compared on this object, which have then been extrapolated to define it's movements through The Milky Way. Around 5 billion years ago, this cluster is believed to have passed right through the disc of the galaxy which portrays a totally new scenario for star formation, other than galaxy collision or supernovas. Astronomers pinpointed the area of contact in the disc at the open cluster NGC 6231 that just happens to be 5 billion years old – so far, so good! The implication here is that this interaction caused gas to be compressed to create the new open cluster, and this scenario could add considerably to the star populations, especially as these collisions could occur every million years or so.

Collisions and mergers within the cores of globular clusters are inevitable. Often, binary stars are created from these collisions, but if one member of the pair is a white dwarf, then a "cataclysmic variable" can be produced. This situation occurs where the dwarf star draws stellar material from its partner star, and if the density reaches a critical point, a sudden temporary brightness can occur. This change is the result of a thermonuclear explosion that astronomers can witness within globular clusters using ultraviolet light.

Globulars are a natural by-product of star and galaxy mergers or collisions and are, therefore, seen throughout the Universe. Astronomers have used massive supercomputers to model the formation and evolution path of globular clusters to display their life and death scenarios in real time. The computer models predict that mergers and collisions are common over long periods of time and that binary stars created under these circumstances will often swap with other member stars.

Globular Cluster Age

Some of the oldest stars in the entire Universe are contained within globular clusters. In fact, some globulars are even older than many galaxies. I have already mentioned the Universe-defying age of some clusters, NGC 5286 in Centaurus being a prime example, aged somewhere between 15 and 17 billion years.

One way to determine the age of a globular cluster is to find the oldest blue star that is still on the main sequence. This gives a reference point as these stars can be more easily aged, as they are still active. Once the age of the oldest globular clusters is determined, this data can also be used to measure the age of the Universe.

One particular cluster, M4 in Scorpius was recently used in studies to date the Universe. M4 was specifically chosen due to its close distance of only 6,800 light-years (it is the Milky Ways closest globular) but it contains extremely ancient stars. This study used white dwarf stars, as these are at the end of their lives so and are some of the oldest stars within the cluster. The age of these white dwarfs can be calculated because they have a known mass, no larger than 1.4 solar masses, and they no longer burn energy because they are in the process of cooling down. The temperature of a cooling star can also yield data that can be used to age it. Astronomers, therefore, use the oldest white dwarfs which appear the faintest and coolest, and the results of this project are now in, which produces a figure of 12–13 billion years for the Universe. Taking into account the time from the creation of the Universe (the Big Bang) to the first stars being formed (about a billion years) gives a final figure for the Universe of 13–14 billion years, which thankfully is very similar to other dates confirmed by measuring galactic expansion rates.

However, not all globulars are quite so ancient, and a recently discovered object in Cetus, called Whiting 1 is believed to be the youngest globular within the Milky Way, at only 5 billion years old. Situated approximately 150,000 light years away in the Galactic halo, Whiting 1 was originally thought to be an open cluster because it only contains about 2,000 stars, although this figure is comparable to the Palomar globulars. Further studies will hopefully determine if Whiting 1 is in fact a captured cluster of extragalactic origin.

Globular Size

The most distant clusters are generally the largest, as they are less affected by tidal forces trying to pull away star cluster members. A typical diameter for a globular is 50 parsecs, or about 160 light-years, but they can be anywhere from several tens to hundreds of light-years across. The stars in a globular are on average packed in at about 1 star per cubic light-year. Similar to the way the Moon and the Sun appear the same size, even though the Sun is much larger, clusters often look similar in size. Most globulars are at least 10 times larger than typical open clusters, but as they are generally at least 10 times further away, globulars can appear similar to open clusters from a visual perspective. Some globulars appear similar at first glance, but they are all slightly different in terms of size, compression and resolved "features" such as chains of stars and dark lanes. By studying the proper motion of globular cluster member stars, astronomers have revealed that certain clusters are travelling at speeds over 1000 miles per second!

Mass

Omega Centauri, the largest and brightest known globular, certainly contains at least a million stars, but is believed to harbour over 10 million, which is comparable to the stellar populations of some small dwarf galaxies. However, this cluster is exceptional and contains ten times more stars than the average globular. In relation to the symmetrical; nature of globular clusters, this is directly related to the distance from the galactic nucleus. Globulars are less symmetric; the closer they get to the central bulge, and clusters further out are more symmetric, as they are less affected by gravitational forces.

Star populations in the galactic halo, other than the globular clusters themselves, are much lower than in the disc and spiral arms where the highest proportion of stars reside. At the clusters cores the stars are much denser, with up to 1000 stars in each parsec (3.26 light-years) of Space. Over a typical globular's life our Galaxy and forces from external sources, play tug-of-war with the clusters member stars. This effect is admirably demonstrated in distorted clusters such as Palomar 5 (Pal 5) which shows chains and strings of stars that have been forced out of the main body. Other small and weak globulars are similarly effected, such as Palomar 10 (Pal 10) in Sagittarius.

Globulars should contain a proportion of gas from the interactions between member stars and the forces from the stellar wind, but until recently, these had not been observed before. Radio Astronomers used 47 Tucanae, the second brightest globular, and detected this gas by studying pulsars contained within the cluster. Pulsars are a form of rapidly rotating neutron star that sends beams of light to us at precise intervals. By timing these signals, astronomers were able to calculate the presence of this gas using dispersion measurements. Comparing this with non-cluster stars reveals 100 times more gas within cluster members, than in unrelated background field stars. However, astronomers believe this gas amount should be much higher, considering the age of these clusters. Ironically, it is probably the pulsars allowing the detection of this gas in the first place, that are responsible for purging its quantities! Binary millisecond pulsars have also been detected in the globular cluster M30 which is currently undergoing core-collapse. The first pulsar, PSR J2140-2310A (M30A) is an eclipsing 11-millisecond pulsar in a 4-hour circular orbit, and its partner, PSR J2140-23B (M30B) is a 13-millisecond pulsar in a highly eccentric orbit. However, the current record holder for millisecond pulsar discoveries is the globular Terzan 5 in Sagittarius, with 24 known pulsars, and at least a further 200 suspected members.

Structure and Composition

Unlike open clusters, globular clusters are gravitationally very stable, some would say infinitely stable, but we can only say this for the cluster ages we see today. This could certainly change in the next 15 billion years. Very few young, blue stars are found within globulars, due to their old age, and most of their member stars are red giants that are off the main sequence, and entering the final stages of their stellar life.

Globulars are generally metal poor, also because they are ancient; they have not benefited from much star regeneration and, so, contain few of the elements created from stars. Little gas or dust is contained within globulars and, therefore, there is not much in the way of star formation within these clusters. Some of the stars in 47 Tucanae have only 25% of the metal content of the Sun, and M15 in Pegasus has almost no metal content at all, with only 0.2% compared to the Sun. Metal content also has a relationship to distance, with close clusters containing more metal and distant objects having lower metal proportions. Omega Centauri, however, has recently been shown to contain two distinct stellar populations of red and blue stars. The blue stars in particular, which make up over 25% of the cluster, are actually higher in metal content than their red counterparts and contain up to 40% Helium, the highest figure found in any cluster so far. In theory, the red stars, which are older, should have a higher metal proportion, but it is thought that this situation has occurred due to the mature red stars turning into super-

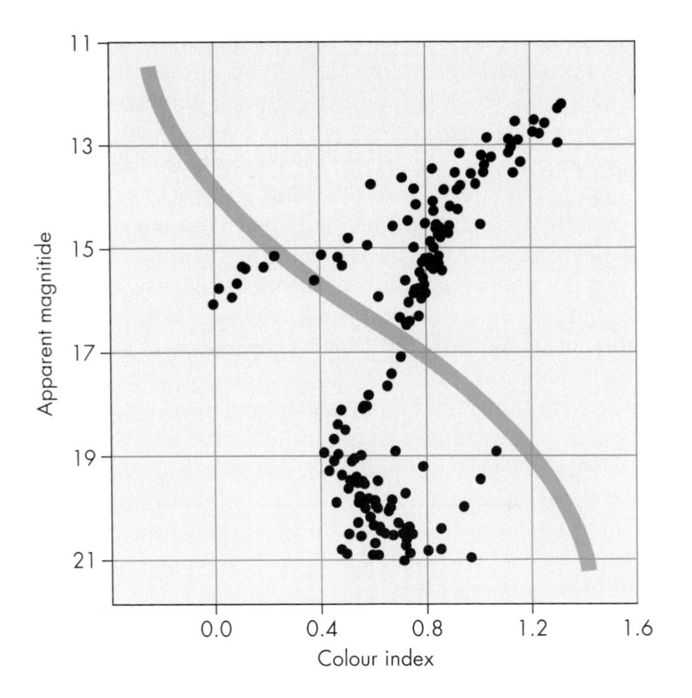

Figure 4.3. A typical globular cluster HR diagram.

novae, which has led to the formation of younger blue stars from this metal rich material.

Globular clusters seem to be contained in two distinct areas: the disc group, which are closer to the galactic center; and the halo group that lie in the outer fringes of our Galaxy. NGC 6352 is a typical disc globular, whereas NGC 1261 is a halo group member.

The Lithium content should also be the same in all stars within a globular, and any differences may be due to varying spin rates of separate stars that may deplete or replete lithium rates. Lithium depletion is used to age clusters, where low mass stars burn their lithium content early in their evolution, so the lithium poor stars are the least massive. Blue straggler (BS) stars can also be found in globular clusters. The theory that BS stars are created when two stars merge, seems very likely when you consider how tightly stars are packed into the core of a globular cluster.

Although many deep-sky objects look monochrome through amateur scopes, several globular clusters do show hints of color, but some of these hues can be caused by optics, recording devices or just plain over enthusiasm. However, subtle yellow and blue shades have been recorded photographically and using CCD imaging.

Origin and Location

Within the halo of the Milky Way, we find many globular clusters of all shapes and sizes, but the halo is still one of the least well-understood areas of the Galaxy. It is known that the clusters are centered around the galactic bulge and reach out for more than 400,000 light-years. Globular clusters helped early astronomers prove that the Sun was not at the center of the Milky Way, but two-thirds out, embedded in one of the spiral arms. By studying the location of globulars, astronomers realized they are not scattered across the sky; in fact, over 30% are concentrated in the direction of Sagittarius, thus defining the galactic center. The remaining clusters are dotted around the halo in a spherical fashion, above and below the galactic disc with an inclined orbit. One strange globular, NGC 3201 in Vela, orbits the Galaxy in a retrograde orbit; in other words, its orbit is in the opposite direction to other star clusters. This is one of several known globulars that move around the Galaxy in this fashion, but there seems to be no clear reason as to why. This particular cluster is notable for its number of RR Lyrae variable stars, but other than this, it is a fairly normal globular cluster.

Recent studies have shown that globulars are not always necessarily associated with a galaxy, which is the standard configuration. Intergalactic clusters have been found near a large galaxy group 400 million light-years from Earth. Although these clusters are near these galaxies, they are not gravitationally bound with them. This is not a new concept for astronomers, but observations have only recently shown this to be the case.

These "wandering" globulars may have been pushed from their parent galaxy, or their original galaxy residence may have been dispersed by interactions with other large galaxies. It is also quite possible that these clusters could find new homes within other galaxies if they got too close and became captured. These intergalactic clusters have not been observed prior to this because of their small size and vast distance.

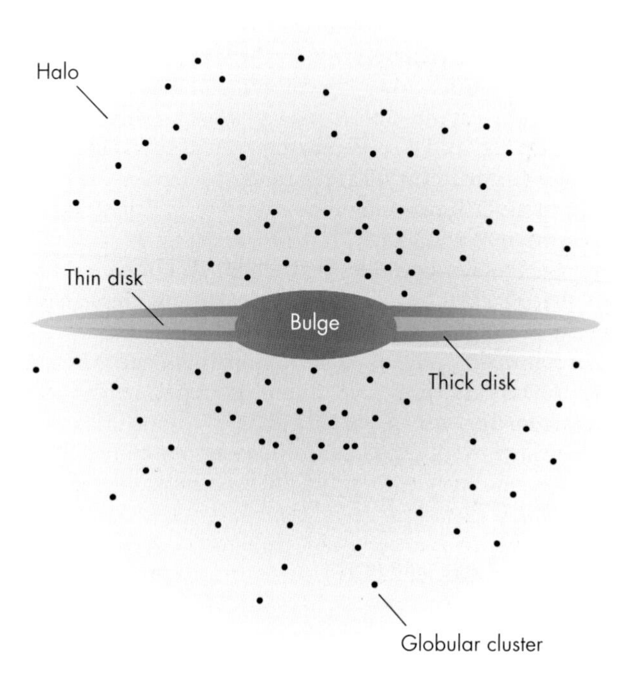

Figure 4.4. Globular cluster distribution in the galaxy.

It would certainly be interesting to live on a planet that formed within a globular cluster. With so many "suns," the night sky would never be dark, and you would not see very much else in the Universe, through this myriad of cluster members. This scenario was thought unlikely, however, as globular cluster stars are old and contain few of the heavy elements necessary for planet formation and, therefore, for life to originate. However, we know of at least one planetary mass around a globular cluster, so its seems likely that other will be found. As with open clusters, globulars seem to group together as clusters of clusters. Some of these are line-of-sight effects, but many are individual clusters that happen to lie in the same area of sky. Ophiuchus for instance contains over 24 globulars.

Distance

Due to the position of globular clusters, far out in the Milky Way's halo, parallax measurement (discussed in Chapter 1) is not an effective method for judging distance. Astronomers can, however, compare with other stars of similar ages and luminosity to calculate the distance of stars within the actual clusters. RR Lyrae stars, a type of variable, pulsating stars normally found in old Population II regions and globular clusters can also be used if present. These are similar to cepheid variables, though much dimmer in luminosity, but they can still be used effectively to estimate the distance of globular clusters. Another type of cepheid variable known as W Virginis stars, which are typically lower in mass and magnitude than standard cepheids are specifically found in globular clusters. Using these methods, however, it is still more difficult than measuring open cluster distances, as globular cluster stars are low in metal and fairly dim, but some useful data has been obtained.

Globular cluster distances range quite dramatically, and are often in the region of 7,000 up to 250,000 light-years away. The Palomar (Pal) and AM (Aaronson, Madore) globulars form part of the extreme halo globulars and, currently, are the most distant. There are around six confirmed extreme halo members. AM-1 in Horolgium is Magnitude 15.8 and 397,000 light-years away, and Pal 4 in Ursa Major is Magnitude 14.2 and 356,000 light-years away. The most distant globular cluster visible in standard amateur telescopes is NGC 2419 in Lynx, which is 274,000 light-years away, and another bright contender is NGC 7006 at 135,000 light-years. Omega Centauri, the brightest globular in the Milky Way, is fairly close at only 17,000 light-years, and within its core the stars have less than a tenth of a light-year between them. By far, the faintest globular cluster in the Milky Way is UKS 1, yet it is surprisingly close, at only 27,000 light-years away. But due to its close proximity to the galactic center, it is obscured in visual extinction and is extremely dim, at a lowly magnitude 17.3.

Some of the most distant globulars have been nicknamed the "extragalactic tramps" as they were thought to wander freely through space. However, a more probable theory to their distance and location is that they formed within the halo originally, and were forced out by interactions with other galaxies. Even though these clusters are at an incredible distance, they are still firmly attached to the galaxy, like a ball on a chain.

Many years ago, the Aricebo radio telescope in Puerto Rico was used to send a signal to M13, the great Hercules globular cluster, in the hope of contacting alien civilizations. (This may seem like ancient history, but there is a valid point to this

Figure 4.5. M13 in Hercules © Cliff Meredith.

Figure 4.6. NGC 2419, a remote globular cluster © Tony O' Sullivan.

story related to stellar distances.) At the time, many thought this was bizarre, but globular clusters are actually a good place to look for life. The stars within them are old, as life needs time to evolve, and their star population is numerous, so the odds of finding that "life-supporting star" are increased. Unfortunately, we are still waiting for a reply. M13 is 24,000 light-years away, so simple maths tells us we will have to wait 48,000 years for "ET" to return our call! The moral of this story (and my point!) is; never underestimate the vastness of Space – no one can really comprehend its infinite size.

Death of Clusters

Computer simulations of the evolution of globular clusters, subject to forces from the Galactic tidal field, demonstrate that binary stars and tidal shock have a massive effect on the life and ultimate fate of globular clusters. Astronomers have probed the "Roche lobes" of these stars, where the gravitational forces of these mutual pairs meet and equalize, forming an hourglass shape around the stars. Contrary to previous theory, clusters on eccentric orbits are vulnerable to strong disk shocks, which dominate their lives until their termination. The destruction of the globular Palomar 5 is predicted when it next crosses the galactic plane in about 110 million years. This timespan corresponds to only 1% of the clusters lifetime, and suggests that many clusters could once have populated the Inner halo of the Milky Way, but have dispersed into fragments from the streams in the Galactic tidal field. It now seems likely that most globular clusters will eventually die, but this may take another 15 billion years.

Classification Systems

Globular clusters are classified using the Shapley system, which is based on Roman Numerals as follows:

From I – XII (1 to 12).
I – Very Compressed
to
XII – Not Concentrated

This is essentially a description of the clusters stellar concentration; however, a clusters spectral type can also be used (i.e., F5), but this is based on a composite measurement and not generally used by amateur astronomers.

New Object Discoveries and Developments

Several new globular clusters have been discovered in the last few years, which include two recent candidates that were found "accidentally" in an automated survey. 2MASS-GC01 and GC02 both reside in Sagittarius, are extremely faint, only 2-3' in extent and are hidden optically but were detected at infrared wavelengths that can penetrate obscuring matter. This is one of the main reasons why so few globulars are catalogued. There is no doubt other candidates exist, they are just hidden from our view. Another recent find is the globular IRAS 18462-0133 which was a previously known, but unclassified object in Aquila that has now been given the tentative designation of GLIMPSE-CO1. This new cluster is somewhere between 10,000 and 17,000 light-years away, which makes it a very close object, and is estimated to be in the region of 3 to 6 light-years across with an approximate age of 10 billion years. Visual magnitude and size data is not yet available, but this new cluster provides further evidence that there must be many more galactic globulars still waiting to be uncovered.

X-ray observations detected from NGC 1851, a globular cluster in Columba, revealed extremely potent emissions that hinted at the possibility of a hidden object of supreme force – a black hole could be lurking somewhere near the center of this cluster. If enough stars merge, then move off the main sequence and turn into supernovas within a short enough time scale, say 20 million years, a black hole can form. However, these are medium-sized black holes, not as large as the supermassive ones found in our own, and other galaxies. Star velocity measurements also show evidence of supermassive black holes in many galaxies, and the mass of the galaxy is directly related to the mass of the black hole. Star velocities in M15 a globular in Pegasus, and G1 an extragalactic globular in the Andromeda Galaxy indicate the presence of black holes with the same size and mass ratios as found in galaxies. The black hole in M15 is calculated to be in the region of 4,000 solar masses, and several young globular clusters have also been observed forming within the galaxy M81 using x-ray analysis, which could also be host to further black holes.

A completely new type of cluster has been found in the lenticular galaxies (between an elliptical and a spiral) NGC 1023 in Perseus, and NGC 3384 in Leo. These so-called "extended clusters" are similar to the globular cluster variety, as they contain very old stars, but they are much larger, typically four times, and also fainter than normal globulars. The strangest property of these objects is the fact they live closer to the disk than the halo, which is a location more suitable to an open cluster, and how these objects formed and evolved is something of a mystery. One theory proposes that interactions between the host galaxy and a close-by dwarf galaxy could have formed a new series of stars that were then pulled near to the disc region, but evolved with a different time scale to the actual disc. Only recently, further examples of extended clusters were discovered in the spiral galaxy, M31 in Andromeda, which are several hundred light-years in diameter. They are also much larger than typical globulars, yet these stars are ancient but hundreds of times less dense than usual globular clusters. The new "extended clusters" are distributed in a spherical halo extending 200,000 light-years from the M31 galaxy and may also have formed from galaxy mergers or interactions. Why these enigmatic clusters are so large, and have been observed in such different environments is unclear, and astronomers still do not understand why no such objects have been found in our own Galaxy, the Milky Way.

Planetary nebulas (the leftovers of a supernova) have also been detected in several separate globular clusters. M15 in Pegasus contains Pease 1, the first planetary nebula discovered in a globular, which is embedded deep within the star cluster. The second discovery, near the core of M22 in Sagittarius, also contains a planetary nebula called JGGC-1, which was recently reobserved at infrared wavelengths to confirm its existence. Originally catalogued as IRAS 18333–2357, the team observed the central star of the nebula and confirmed the data corresponds to the optically achieved results. Two further nebulas were detected in a survey of 133 galactic globular clusters conducted exclusively to find nebulas within globulars. The objects discovered were found within the globulars Palomar 6 in Ophiuchus and NGC 6441 in Scorpius. These new objects are catalogued as JaFu1 (Pal 6) and JaFu2 (NGC 6441).

During a survey of M87, an elliptical galaxy in Virgo, astronomers have also detected the first modern discovery of a "classic" nova (a star, which suddenly brightens, then recedes) within a globular cluster. The last definite classical nova eruption was nova T Sco in M80, way back in 1860! This time around, the erupting stars light curve and color matched those expected for a nova at the distance of M87. Astronomers cannot rule out the possibility that the nova is not directly related to the cluster, but their data proves it is very unlikely, as the globular cluster nova frequency is much higher than previous observations have suggested.

The first planet within a globular cluster was recently discovered within M4 in Scorpius, the closest globular with an age of 12.6 billion years old. The extrasolar planet was detected orbiting a millisecond pulsar, B1620–26, with a third object, a white dwarf star companion also within its orbit. By combining the measurements of this triple system, a planetary mass was defined of 2.5 Jupiter masses, orbiting about 25 AU from the pulsar. This discovery defies earlier theory that planets cannot form in metal-poor regions, such as globular clusters, as there are no heavy elements for them to originate. However, only a few years before this study, another team searched 34,000 stars within the 47 Tucanae cluster for extrasolar planets and found none. These findings challenge both star formation and planet formation theories, as planets should require gas giant stars to promote growth, and gravitationally stable zones to evolve planets; neither of these occurs in a globular cluster. Ultimately, it is possible that planets were formed much earlier than we thought possible, and that other globular clusters may, in fact, be host to planet formation.

Summary

The study of globular clusters has been a particularly active area of research in the last few years, and there have been some major discoveries and surprises for professional and amateur astronomers alike. What were once thought to be immense sleeping giants, lying dormant for millions of years, are now known to be active, evolving systems that are truly alive and contain astronomical objects that previous generations would not believe or comprehend. The relationship between globular clusters, galaxies, and the Universe as a whole is slowly being unravelled and the tales of mergers, collisions, cannibalism and star formation may only be the start of this story.

Chapter 5

Stellar Associations

No matter where we look in the Universe, we find that stars always form groups. Within our local solar vicinity, these groups take on the form of stellar associations, and as they are only recently being studied in any depth, our knowledge of these systems is not as comprehensive as it is for star clusters. In structure, these groups are similar to open clusters, but they are much larger, loose groups of low-density stars often associated with smaller open cluster groups. The components of these associations are recently formed stars that were born primarily in the spiral arms and are loosely bound together. Observations show that associations are nearly always found buried within giant molecular clouds, which implies that this material created these stellar groupings. The member stars are all the same ages and must have been born at the same time from the same raw material. After these groupings have formed, the gas and dust is eventually removed, and the association is left behind. On an average, the associations contain between ten and a hundred stars, but a few candidates such as Cygnus OB-2 contain up to a thousand members, or more. Although physically attracted, associations are not strongly bound by gravitational forces; so they are fairly short-lived phenomena lasting on average only a few million years. Eventually, the association stars around an open cluster will be dispersed, leaving only the actual cluster still visible. There are roughly 70 known examples of stellar associations, and many of these are readily visible through amateur telescopes. Astronomers were able to confirm these groups because their member stars are concentrated in the same part of space, and they are moving in the same direction. Associations are known to be a common birthplace for many stars and are important "test beds" for star formation and evolution theory.

To identify associations, star counts and spectral classification is used to detect concentrations of stars that have higher densities than the background stars. Astronomers use a grid system to compare suspect clusters to those with known quantities; they then search for bright magnitude stars with O and B spectral types. There are three classifications for stellar associations, bound, intermediate and unbound, but there is not always a clear distinguishing line between associations and clusters. To determine a cluster is fairly easy, as they are more easily recognized; however, associations are often difficult to detect because they have such a large surface area, which blends in with background and foreground stars. Currently, there are two basic types of association, OB or O Associations, and T Associations with an average size of 260 light-years. It is not known for definite whether these types are completely separate regions of star formation, or if they are connected in some way.

OB Associations

O or OB associations, as the name suggests, are groups of massive hot stars with spectral types O and B that are typically spread over an area several hundred light-years in diameter. Their component stars are very luminous. OB associations can be found centered around several prominent open clusters, including the Double Cluster in Perseus and within the Orion Nebula (M42). Their ages span from a million to perhaps 30 million years old, and they play an important role in the theory and observation of protoplanetary disk formation. This field of study is very active, and new discoveries are being made in a rapid fashion. For example, it is now known that associations contain dust and gas discs that show clear indications of disc evolution, and that emissions from these discs decrease with age.

Extragalactic OB associations are found in other galaxies, including the Large Magallenic Cloud, a satellite galaxy of the Milky Way. Many such associations have also been found in compact groups residing within the Andromeda Galaxy, M31. The average size of the OB associations in M31 is 250 light-years, which compares well with the sizes determined in the LMC and in the Orion OB association in our own galaxy. Large groups of associations in the Andromeda Galaxy contain many Cepheid variable stars that also help to determine the stellar distances for these associations.

The Upper Scorpius OB association is one of the closest star formation regions we can readily observe today. At a distance of only 470 light-years, and an age of around 5 million years, x-ray observations have studied 30 high-mass stars known to reside within the group. Recent studies have identified three subgroups related to the association, Upper Scorpius, Upper Centaurus-Lupus, and Lower Centaurus Crux. A hot circular bubble of gas has been observed containing these associations, and over 100 low mass members have been studied to determine their age and mass. Stars began forming in these regions about 15 million years ago, but there have been several other bursts of star formation since this period, with successions of supernova outbursts between them. The most recent formations commenced approximately 1 million years ago, and new findings suggest the Scorpius association has undergone at least 20 supernovae explosions in the past 11 million years, and Antares, a massive bright yellow star, is likely to be the next victim.

The OB associations that we observe today may be the cores of older groups that have long since vanished to the rest of Space, but there does appear to be some form of hierarchical system for associations of stars, and by implication, star formation. Many OBs contain smaller groups, or subgroups, such as the Double Cluster, and these subgroups are made up from multiple stars themselves.

Some of the most prominent OB associations are listed below:

Auriga has two associations, Aur OB1 and Aur OB2, that are very close to one another.

Cassiopeia contains six prominent OB associations, Cas OB1, Cas OB4, Cas OB5, Cas OB 6, Cas OB 8 and Cas OB14. All these groups are situated in or near the constellations main "W" or "M" asterism.

Cepheus has 4 OB associations, Cep OB1, OB2, OB3, and OB4. The brightest member is Cep OB2, situated near the star nu Cephei.

Cygnus also has six associations, Cyg OB1, 2, 3, 4, 7 and 9 and contains a star (No 12) that would be one of the brightest in the sky at absolute magnitude –9.9, but interstellar absorption dims it to a lowly magnitude of 11.5! The Cygnus OB-2 association has recently been found to be emitting extremely strong unidentified gamma rays, probably caused by stellar winds from interaction within the young cluster stars. This could be a source of the elusive cosmic rays, but further studies will be required to confirm this. Cyg OB-2 is also the largest stellar association in the galaxy, and contains at least 2,600 OB stars.

Gemini contains only one OB association, Gem OB1, near the open clusters M35 and Collinder 89.

Orion contains four related associations, OB1-A, B, C and D.

OB1-B surrounds the belt of Orion, OB1-A covers an area northWest of the belt, OB1-C contains the stars around the sword, and OB1-D is centered on the Orion Nebula.

Other objects in the vicinity include NGC 1981 and Collinder 70 open clusters, and the Orion OB association has ages from 3–10 million years obtained from recent studies.

Perseus has two main associations, with Per OB3 around the bright star Alpha Persei, and Per OB2, which is the source of a "runaway" star Xi Persei.

Scutum has only one stellar association, Sct OB1, which is directly related to M11, the Wild Duck Cluster.

T Associations

T associations are similar in nature to OB associations but they contain low mass, T Tauri type stars. T Tauri stars are newborn stars, or protostars, that are still evolving and undergoing contraction. They are variable stars with an irregular period, and are nearly always found in groups or associations, with temperatures ranging from 3,500 to 7,000 Kelvin (K). Most T Tauri stars are found within dense interstellar clouds and can be detected within other associations and groups. For example, there are estimated to be over 2,000 T Tauri stars in the Sco-Cen OB association, and they have also been found in the open cluster Collinder 197. T associations also reside within the Vela OB association, centered on the star gamma velorum. These low-mass, young stars radiate strong xray emissions, but there has been no evidence for a debris disc that could indicate protoplanetary formation. Its T associations, T1 and T2 contain about 200 T Tauri stars, within the OB association that lies at a distance of 410 pc.

T Tauri stars are surrounded by an envelope of gas that produces double jets of streaming gas (bipolar outflows) that emit strong infrared radiation and travel at several hundred kilometres per second, and they can last about 10 million years.

In Chameleon and Lupus, T associations have been observed with physically bound stars, binary stars and multiple systems. Both the Chameleon and Lupus associations provide an interesting pair to study, as they lie at similar distances of 554 light-years, and 489 light-years respectively. Binaries are common in these associations, and many were formed early in the association's life; however, their numbers remain fairly constant throughout the group's lifetime. These binary stars were probably created from collisions within the member group from lone stars, or other multiple star systems.

The recently discovered Horolgium Association, which is only 200 light-years away, contains over a dozen young stars, half of which are also in the T Tauri class. Also known as the Horolgium/Tucana Association, this group is one of the youngest, at 20–30 million years old. Follow up observations discovered 11 new member stars, only 50–60 pc from Earth with these being the most northerly members detected so far.

The closest known T association in the Milky Way is the TW Hydrae association, where a number of new T Tauri stars have been observed. Eleven systems containing nineteen new young stars were discovered, betrayed by their high x-ray emissions, strong activity in their gaseous outer layer (chromosphere) and increased lithium content. All of the stars in the study have the same proper motion, and taking into account all of the data, confirmation of their identity as genuine T associations was determined. One of the stars, an M5 type, is less than 22 pc from Earth, which makes it a realistic candidate for planetary detection. From this survey, astronomers were able to calculate the age of this association to be in the region of 8–10 million years, with a distance of 163 light-years. The age and proximity of this group also makes it an ideal target for the study of brown dwarf stars, and the possibility of planet formation. However, the most fascinating discovery was the fact that the TW Hydrae grouping is not associated with any known gas cloud, which was the first example of its kind.

The "Star Complex"

Star complexes can be considered the largest physical structures of stellar groups and associations, and encompass star regions up to 1000 parsecs in size and stars up to approximately 100 million years old. Initially, these massive star cities were observed as clumps in the spiral arms of extragalactic galaxies. They contain the largest and oldest groups of young clusters, from binary and multiple stars to associations and super associations, as they were originally named. The youngest and smallest clusters are always found within older and larger systems, so some form of hierarchy is certainly present. Many galaxies are known to contain at least one or two star complex regions, with bright OB associations, and it is these complexes that have helped to define the structure of the spiral arms, in our own galaxy and other external galaxies. Bright complexes that are large and blue are located in the inner arms of galaxies, and are known regions of star formation, called starburst clumps, although older complexes appear to trace the outer arms of the spirals. It has been suggested that over 90% of OB Associations and other young clusters in the Milky Way, Andromeda Galaxy, M33 Galaxy and the LMC are in fact united as massive star complexes. These extremely large star groups have been found by observing cepheid stars in regions up to 1,800 light-years in diameter. Within the LMC, there are over a dozen cluster clumps in NGC 2164 all of which have similar ages. The ages of these vast systems are believed to be up to 50 million years old, with one region in the Milky Way that encompasses several separate stars forming regions. This system, known as "Goulds Belt", includes the Orion, Perseus and Scorpius-Centaurus OB associations. Older associations of stars are more scattered than their younger counterparts, such as the Cassiopeia-Taurus association, where star formation ended approximately 20 million years ago.

Moving Clusters

Moving clusters are gravitationally the weakest associations of stars. They are collections of unrelated stars, but have common motions and velocities, moving through Space in almost parallel paths. From the Earth's viewpoint, the stars in a moving cluster appear to be travelling towards the same point in the sky, called the convergence point. This effect is purely down to perspective, but the parallel tracks that these stars follow appear to meet at point away from the clusters original location. The position of this point can be measured from the angles of these tracks, using proper motion studies, and from these measurements, the clusters distance can be calculated. Astronomers have used also these methods to determine the distances of the Ursa Major Group and the Hyades cluster in Taurus.

The Ursa Major Group is one of the best-known and most-studied moving clusters and includes the bright stars of Ursa Major (within the Plough), and the most luminous members of other constellations, including Leo, Canis Major, Auriga and Eridanus.

The Sun is also believed to be a member of this moving cluster, so the clusters age must be around 4.5 billion years old, assuming it formed from the same material, and at the same time as the Sun. Moving clusters could account for up to 10% of all the stars that have catalogued velocity measurements.

Another moving group, Beta Pictoris, which is 12 million years old, was probably created from supernova outbursts in the Scorpius-Centarus OB association, and from observation it was determined that the group probably formed unbound. 11 new member stars, including five binaries, have been discovered recently within the group, with many of these situated in the Northern Hemisphere. This group contains a variety of spectral types and its longest dimension is believed to be stretched out over 100 pc, whilst a protoplanetary disc is evident from observations.

Summary

Stellar associations are full of surprises, they are active star forming regions that continue to excite and puzzle professional astronomers. Their relationship with brown dwarf stars and their companions is at the forefront of research, and these objects are being found in many associations. Discs of debris that may become planet-forming discs are common enough in stellar associations, that astronomers have been able to formulate a model. It appears that a 10 Jupiter mass planet at a distance of 100 AU from the parent star can be considered a standard.

Chapter 6

Asterisms

In reality, asterisms are not true star clusters at all; they are star groups or patterns, and are not generally physically related. Just like the constellations, where many asterisms play a prominent role, these objects are mostly line-of-sight effects, or optical illusions. Their member stars are all at different distances but perspective allows them to line up in the sky, making interesting and memorable shapes. This is not to say that an open cluster cannot be called an asterism, because by definition an asterism is just a pattern of stars, but clusters are always related stars, so it may be easier to think of an asterism as a non associated group. Many asterisms have been confused in this way, as genuine clusters throughout the ages, so they deserve a place in any star cluster discussion.

The patterns of the constellations have both ancient and modern symbolic interpretations, for example the constellation Ursa Major is the "Great Bear" but it is better known for its asterism, "The Plough" or the "Big Dipper." This contains seven stars, four, of which make up the "bowl" and three the "handle." It is actually often easier to find a constellation from its asterism rather than its official constellation outline, such as the famous "Water Jar" in Aquarius, which makes a "Y" shape with the stars gamma, eta, pi and zeta Aquarii. Asterisms also provide a useful tool for navigating the sky using pointers, for instance, the outer two stars on the bowl of the "Plough" point to the Pole Star, which is often required for polar aligning an equatorial telescope mount. These pointer stars also point down to the star Regulus in Leo, and the "handle" stars point in the direction of Arcturus in Bootes – so asterisms have a hidden talent! Similarly, the belt stars of Orion (also a famous asterism) point up to Aldebaran in Taurus, and point down to Sirius in Canis Major. These are only a small selection of the built-in star pointers of many asterisms, but if you know a few, you can negotiate most of the celestial sky.

Asterisms are not officially recognized by the International Astronomical Union (IAU), the world-wide astronomy governing body, but they are widely recognized and used by amateur astronomers (and a few professionals) throughout the globe.

What could be considered the original or "classic" asterisms, were based on old constellations that no longer exist, many of which originated before the IAU drew up official boundaries for the 88 constellations we recognize today. Some of these older asterisms were also based on constellations from other countries and cultures, such as the Chinese, who devised their own system to navigate the night sky. Examples include Officina Typographica, Noctura, Triangulum Minor and Felis, that can all still be seen in the sky (if you use an old star map) but have no real place in modern astronomy. Star groups such as the Pleiades and Hyades could also fall into this category, as they have been recognized for centuries, but modern astronomers class these objects as open clusters, not asterisms, because they are not juxtaposed stars, but physically bound groups.

Most of the well-known asterisms are associated with the component stars of a single constellation or part of a constellation, such as the "Keystone" in Hercules which are the four bright stars that make up the "heroes" body. But several separate constellations can be involved, for example the "Square of Pegasus" is again an association of four bright stars, but some of these stars actually belong to another constellation, Andromeda. There are many other examples of large bright asterisms that are based on subconstellations and are mainly visible with the naked eye, or binoculars. These include, the "W" shape in Cassiopiea, the "Sickle" in Leo made up from the stars Regulus, eta, Algieba, Adhafera, Rasalas, Algenubi and Alterf, and the "Northern Cross" in Cygnus. The "Circlet of Pisces" is also an attractive asterism, formed by seven stars that define a ring shape. The member stars are gamma, beta, theta, iota, lambda, kappa and 19 Piscium. Another famous example is the "teapot" asterism of Sagittarius, which is very obvious and easily seen with the naked eye in the Southern Hemisphere.

Even larger asterisms can be seen at particular times of the year, which are based on bright stars from various constellations. The "Summer Triangle," which incorporates the stars Vega, Deneb, and Altair, and the "Winter Triangle" that is made up from three stars in Orion, Canis Major and Canis Minor are difficult to miss! Though these large and obvious asterisms are interesting and fun to observe, for the scope of this book we are more interested in dealing with the more compact asterisms. These are the objects that resemble, or can be confused with open or even globular clusters, such as Kemble's Cascade and the Coathanger or Brocchi's cluster.

Kemble's Cascade is a smattering of stars near the constellation Camelopardalis and contains about 20 stars that form a "string of pearls" that covers over 2.5° of sky, or nearly five times the width the moon. It is fairly close to a compact open cluster, NGC 1502, and is a very admirable object for a "mere" asterism. The Coathanger, or Brocchi's cluster is another splendid object, that was originally thought to be an open cluster (I listed it as such when I first observed it!) but is actually an asterism in Vulpecula, close to the Sagitta border near the globular cluster M71. It looks very much like its namesake, but the stars are not related, just an accidental group of "arranged" stars. In fact, its member stars range in distance from around 200 light-years, to over 1,200 light-years, so this really is just a chance alignment.

A further "class" of asterism that is also very "cluster like" has been coined the "mini asterism," which generally requires optical aid to observe. These collections of unrelated stars are often discovered and named by amateur astronomers, but there is a large number available in general catalogues such as the NGC. Although asterisms are technically not associated stars, some of the popular ones are actually open clusters, but are also referred to as asterisms because they resemble some form of object. Famous examples include the Jewel Box, NGC 4755 in Crux, and Melotte 111 situated in Coma Berenices. Some of the newly catalogued mini-asterisms include the "Little Queen" in Cassiopiea, and "Dolphins Diamonds" in Delphinus, both discovered by UK amateur and author, Phil Harrington. Southern objects that have received their own designations, include the "Golden Snake" in Circinus, found by Andrew James, and the "Mini Coathanger," discovered by Tom Whiting.

Another "asterism hunter" Yann Pothier is responsible for recognizing at least 17 objects that have been dubbed with the reference ANR (Amas Non Repertoires) which means "uncatalogued cluster." A typical example from Yann's list is ANR

1947 + 18, where the numerals refer to the Right Ascension and Declination positions of the object. Yann also discovered a new asterism in M73, which is called the "M73 Aquarius Asterism."

A extra addition is "Markov 1" an impressive asterism in Hercules, discovered by Paul Markov, whilst observing the DoDz (Dolidze) clusters, near the star xi Herculis. Its shape is reminiscent of the "teapot" in Sagittarius, and detective work produced by Markov rules out the possibility that the stars are related.

In the NGC and Revised NGC alone, I found over 100 objects listed, that are thought to be asterisms, but were at one stage believed to be related stars within a cluster.

Early versions of Norton's Star Atlas also cite over 150 asterisms and candidates.

When these objects were originally discovered, motion and velocity studies, and to some extent, spectroscopy was not available or not advanced enough to show that many of these "clusters" were actually just random groups of totally separate stars. Other observers were simply mistaken when observing these items. It should be emphasized that although these star groups are not genuine clusters, many of these asterisms resemble true clusters and are worthy targets for observation in their own right. In Chapter 9, I will discuss in more detail, some of the misidentified and nonexistent clusters that have been catalogued over the years.

Examples of more obscure asterisms include, NGC 1790 a group of 8 or 9 stars in Auriga, NGC 2409 in Puppis, which also contains 6 to 8 stars, and NGC 6863 a small group of stars in Aquila. The RNGC catalogue also contains many objects that are simply collections of two to five stars that are similar to, but not really classed as asterisms or clusters. There are lots of other asterisms that can be tracked down, and many of them have strange names, often penned after their discoverers, but as they have no official meaning this is not really a problem. Objects include the "Northern Fly" in Aries, "Job's Coffin" in Delphinus and the "Bull of Poniatowski" in Taurus, which you may wish to look up on your next observing session. From the many thousands of stars visible in small groups there is a wealth of "new" asterisms just waiting to be found and named, and although there is no fame or fortune to be had from this venture, there is certainly no harm to be done from pursuing this activity. It is best performed from a visual, observational perspective, rather than by studying a sky atlas and you should be more concerned with new, interesting uncatalogued groups, rather than renaming or reforming the existing ones.

Chapter 7

Extragalactic Clusters

It is becoming increasingly obvious to astronomers throughout the world that the Universe repeats itself wherever you care to look. Stars appear as groups, collate into clusters, which then form associations and these, in turn, reside in massive star complexes. All of these objects occur within a single galaxy, and looking beyond our own Milky Way into extragalactic Space, we see these same repetitions of scale and size duplicated throughout the many observable galaxies. Even the galaxies themselves occur in clusters, and super clusters, and there seems to be no end to this structure, going way back to the very beginning of time when the Universe itself was born. The Milky Way is certainly not unique in this aspect and so we can observe many globular, open clusters and stellar associations in other galaxies, as well as young massive clusters, and super star clusters. These are the extragalactic clusters that will be discussed in this chapter.

Extragalactic Globular Clusters

In our own galaxy, it is relatively easy to resolve stars within open and globular clusters, but in external galaxies due to the tremendous distances involved this becomes much more difficult. However, unlike lone stars at extragalactic distances, the collective stars of clusters, especially globulars can be seen far beyond the Local Group of galaxies. Because globular clusters are relatively simple stellar groups, containing stars of similar ages and composition, they are often easier to understand than the galaxies they reside in, which contain a chaotic mixture of stellar types, all at different stages of evolution.

Almost every galaxy so far observed (except the smallest dwarf galaxies) contains globular clusters in some form. Their numbers range from a couple of clusters in dwarfs to over 10,000 cluster members in the largest cD galaxies such as NGC 1999 and M87. The giant elliptical galaxy M 87 has at least 4,000 confirmed globular clusters, but this number could, in fact, be over 15,000.

The field of extragalactic globular research is currently very active, and new discoveries are plentiful. Very young massive compact objects, which are still forming and evolving, have been discovered embedded within rich gas regions of interacting and merging galaxies, from which new cluster formations are being created. It appears that these mergers result in two separate populations of globular clusters, the original clusters from the parent galaxies and groups of new clusters formed as a result of these mergers. These findings shed obvious doubt on the original theory that all globulars are old objects that formed first within any given galaxy. Optical studies of these merging galaxies show blue and red populations of clusters within elliptical galaxies, but it is not clear whether this color demonstrates differences in age, chemical properties or a combination of the two.

Further research on infrared wavelengths has used spectroscopy to define the ages and chemical content of these clusters, and has shown these blue and red populations to be only currently found in bright, high mass galaxies. In elliptical galaxies, the red globulars are found within the galaxies bulge, and the blue globulars are located in the galactic halo, and their metal content is lower than those found in spiral galaxies. Red globulars are thought to have formed at the same time as the galaxy itself, but the blue population formed at a later epoch, possibly from mergers, as noted earlier.

These compact clusters have also been observed in starburst galaxies as also several normal spiral galaxies, and astronomers have witnessed the clusters in the process of removing the dust and gas from spherical envelopes that contain the newly formed stars, which appear as columns of clouds. Contained in these clouds, or HII regions, are hundreds of young massive stars similar to those in the Milky Way, although they are much larger in extragalactic clusters. Certain regions where star formation is even more productive are known as Ultra Dense HII regions, and these may be the birthplaces of super star clusters (see section on SSC's), but they are not found in the Milky Way galaxy. The earliest stages of star formation can be examined in these regions, but because of distance, Local Group galaxies such as the Magallenic clouds (LMC & SMC) are probed using radio wavelengths to delve though the gas and dust in these regions, that is often obscured in the optical sense.

It is widely believed that these young massive objects are actually globular clusters in the process of formation, and so they have been termed proto-globulars, but it is unlikely that all these clusters will reach this eventuality. However, it is fairly certain that some compact clusters will survive to form globular clusters; for example, the galaxies NGC 7252 and NGC 3921 contain clusters that are already in the region of 500 million years old. Their size and mass are comparable to, or slightly larger than the Milky Way globulars, and their ages range from 1 to 500 million years old. At least 5,500 young compact clusters have been detected in extragalactic galaxies to date, aged between 3 million and 1 billion years old, but this number will only increase as more data comes in. From a research point of view, this is a unique opportunity to witness globular clusters forming today, as opposed to trying to work out how old clusters formed 13 billion years ago, when the Universe was first born.

Formation studies of globulars reveals information on how their parent galaxies formed, but because Milky Way globulars are old, and stripped of their original stellar material through their evolution, their initial matter has long since scattered. However, the LMC clusters are still young enough for their stellar material to be preserved. Metal-poor globulars appear to form at the greatest distances in a galaxies early life and metal-rich globular clusters increase in numbers from mergers and interactions. The cores or central nucleus of these objects are very dense, so technology is only just allowing these areas to be sampled. More data is required on high-resolution spectra of globulars, component stars, and their chemical composition, but these are difficult to determine with current equipment. Therefore, many errors and uncertainties still exist in relation to cluster distances, but new technology and telescopes in the next decade will improve this with better parallax measurements for globular clusters. Color magnitude diagrams and spectra for extragalactic globulars are not currently possible, but will no doubt happen in the future, with even larger telescopes currently being planned.

Only recently, a group of newly discovered stars called Willman 1 were uncovered whilst astronomers were searching for faint satellite galaxies within the Milky Way. This new object is extremely faint, less luminous than any known galaxy previously observed and may be some form of exotic globular cluster, although it is fainter and less compact than typical globulars.

Computing power has also improved the ability to model some of these massive star-forming regions in external galaxies, and spectroscopic analyses of extragalactic globular clusters at optical and infrared wavelengths have been used to measure the ages of globular systems. One question that remains unanswered is: why do globular clusters appear very similar, in very different galaxies? The answer may simply be that globular cluster formation takes on a similar role in all galaxies.

Extragalactic Open Clusters

The open clusters in our Galaxy do not appear to be represented in the same way in other galaxies. However, the young massive star clusters in extragalactic galaxies are not generally found in the Milky Way either. This may be due, in part, to the fact that we have an "inside-out" view of open clusters in our Galaxy, they are also much closer to us and, therefore, more easily detected. It has also been suggested that massive star clusters may simply be a natural progression from normal open clusters rather than a completely different type of cluster. Again, this may boil down to location, location, location! If the Milky Way did not have enough stellar material when it was originally created, perhaps these massive clusters could not form, or maybe our Galaxies initial star formation age was an inefficient one.

M31, the Andromeda Galaxy is famous for the number of observed open clusters it contains, which are mainly old and intermediate age clusters. Compared to the open clusters in the other nearby galaxies, M31 contains less young massive clusters, but they are similar in luminosity and spectral type. M31 also has less red, bright clusters, but contains a similar amount of open clusters as the galaxy IC 1613 and the LMC. Over 230 open clusters have been catalogued in M31, and their structure has been determined using photometric measurements. Open clusters in the inner arms of the disk, 6 kpc from the center, appear redder and are, therefore, older than those found in rich star formation regions 10–12 kpc from the central nucleus. C107 is a prominent open cluster within M31, but these open clusters will typically have a relatively short lifespan.

Two previously known open clusters in the recently discovered Canis Major Galaxy are now believed to be associated with this galaxy. AM-2 (Arp Madore) and Tombaugh 2 are both old open clusters and their inclusion within this galaxy was determined from position distances and from stellar populations in the area. The Canis Major Galaxy is also a new contender for the closest galaxy to the Milky Way at approximately 26,000 light-years. Unfortunately, wide field studies of extragalactic open clusters are few and far between, mainly due to the fact that background stars interfere with the possibility of any new findings.

Extragalactic Stellar Associations

These star groups have not been studied in as much detail as other clusters at an extragalactic level, but they certainly do exist. The galaxy IC 1613, a dwarf irregular type which is a Local Group galaxy situated in Cetus, has been noted for its star formation regions. This site is rich in HII regions (these are areas of ionized hot gas surrounding young stars) and contains at least 20 known stellar associations, with another possible 43 star cluster candidates. There are, however, very few observations of OB associations in extragalactic objects, no doubt largely due to the difficulty in resolving these hot young stars at extragalactic distances, though new technology will improve this situation somewhat.

The observations for extragalactic associations are based on data and sampling techniques for galactic OB associations, and several existing catalogues have been reviewed to find possible new objects. M33, the Pinwheel galaxy and NGC 6822 (Barnard's Galaxy) in Sagittarius have revealed many new stellar association possibilities. A region of 22 pc was examined around M33 and 48 associations were discovered. Similarly, an area around NGC 6822 of approximately 40 pc was studied to uncover at least 3 new stellar associations. These samples also suggest that at distances up to 22 pc from the galactic center, M33 and our own Galaxy have similar star association members, but in M33, as the search area widens, the number of associations dramatically increases. The surface density of M33 is, however, higher than in the Milky Way, so this could account for the increase in association populations.

Star Clusters in Starburst Galaxies

Starburst galaxies have also become a recent target for the observation of new cluster formation, though they are less frequent than in galaxy mergers. This means that young star clusters can originate in a variety of very different environments. Luminous young clusters seem to be a common component wherever there is massive star formation occurring. As much as 20% of the luminosity emanating from starburst galaxies is coming directly from the clusters contained within them, but in our Galaxy, the Milky Way, this number is only 0.01%, demonstrating the intense activity in these regions. It is also possible that many of the field stars in starburst galaxies were originally cluster members that have been dispersed through the ages, so this figure could be much higher. One recent theory infers that all the stars in these galaxies were originally formed within these massive star clusters.

Super Star Clusters and Young Massive Clusters

"Super star clusters" (SSC) are young, very luminous and compact, with up to 100 times more activity than stellar associations and open clusters. They are formed mainly but not exclusively from galaxy mergers and several SSC show a type of structure that evidently point to these events. Starburst galaxies are also known to contain super star clusters, and there are at least 80 known examples in the Local Group of galaxies, such as M82. SSCs can contain several hundred thousand stars, and are host to the most concentrated star-forming regions other than galactic cores. They are formed in high-pressure regions, and provide insight into the conditions of newly formed stars. Initially, astronomers believed super star clusters only existed in groups, and at large distances within merging galaxies. However, a new group has recently been discovered only 10,000 light-years from our Galaxy. The new SSC, called Westerlund 1, is obscured by dust and gas, making it difficult to detect, but astronomers believe it is the closest, and most massive, compact star cluster currently known. It contains hundreds of high-mass stars, with over 100,000 times more mass than the Sun, and has a diameter of about 6 light-years. Due to its close proximity, Westerlund 1 is technically a Milky Way cluster, so it is one of the first SSC's to be found in our own Galaxy, although it is listed in the latest catalogue as an open cluster.

"Young massive clusters" (YMC) are also found in similar regions to super star clusters but they pose a problem to astronomers; they are massive, like globulars, but also young like open clusters. Astronomers are, therefore, trying to find a common thread for these different types of external clusters to find how they relate to one another, because the traditional classification and relationship between open and globular clusters appears somewhat invalid beyond the Milky Way system.

It has been proposed that super star clusters, and another form of expansive cluster known as "scaled OB associations" (or super association), are actually different types of young massive cluster. This would mean that the super star cluster is actually a subgroup classification. Scaled OB associations are extended in size, but lack any determinable structure and are not generally associated with haloes, although they are unbound, short-lived systems. Whereas, super star clusters have compact cores, show some structure and are often associated with haloes. They are gravitationally bound, and long-lived systems that may eventually form globular clusters.

Certain scientists have however, objected to the term "super star cluster" and prefer the "young massive cluster" designation, as it is a more accurate description. Following this example, all young clusters can be classified as an open cluster, OB association, scaled OB association or a super star cluster (which is a type of young massive star cluster using the above system). All old clusters would, therefore, simply be referred to as globulars.

It has even been proposed that all of these extragalactic cluster types are just variations on a theme, no different from the open, globular and association types we see in our own Galaxy. Perhaps they are just at different stages of evolution in contrast to the Milky Way, and because they are located in a completely distinct environment, they cannot be easily compared.

Galaxies With Notable Star Clusters

The Antenna Galaxies

Resolving stars within extragalactic clusters is difficult, but there is one notable exception to this rule. Within the gaseous clouds of the dual Antenna galaxies NGC 4038 and NGC 4039, astronomers have witnessed over 1000 clusters forming, as these two galaxies are colliding and merging. Many of these are super star clusters, and typical cluster ages in this region are between 1–30 million years old. Whilst current theory predicts that globular clusters were created before most galaxies and are, therefore, the oldest "swingers in town", this scenario gives further reason to believe that not all globular clusters are ancient, and so their age is dependent on how and where they were created.

M33

M33, the spiral galaxy, is also part of the Local Group, and is a nearby, almost face on galaxy that provides a good host to study clusters. There are over 60 confirmed globular cluster candidates in this galaxy, with 20 definite objects, C39 being the brightest example. Many of these are small, old globular clusters with distances in the region of 840 kpc. There are at least 248 other clusters within this galaxy, and many more certainly exist. A new survey detected 102 star clusters in M 33, using deep-field imaging, and 82 of these clusters were previously unknown. Photometry measurements reveal that 25 of the clusters have emissions, implying they are globular clusters, and 15 of these are completely new discoveries. This increases the previous globular cluster population by approximately 60%. The new globulars are aged between 6–15 billion years, and their proposed masses range between 100 and 100,000 solar masses.

Sagittarius Dwarf Galaxy

The story of the globular cluster M54 is quite interesting and deserves some explanation. It is actually a member of the Sagittarius dwarf galaxy, so you would assume it to be an extragalactic cluster. However, this galaxy is in the process of being consumed by our galaxy, so it is technically part of the Milky Way and, therefore, not strictly extragalactic. The M54 globular cluster itself is also believed to be the remnant core of the Sagittarius Dwarf galaxy, so you may find it listed in galactic and extragalactic catalogues, but essentially, they are both correct! Other Sagittarius Dwarf globular clusters include Arp 2, Terzan 7 and Terzan 8.

Large and Small Magallenic Clouds (LMC & SMC)

The LMC contains at least 2,000 globular and open clusters, which have a mixture of ages from young clusters and associations less than 1 million years old, intermediate clusters around 5 billion years old, and old globular clusters up to 13 billion years old.

It is certainly evident that there are two distinct systems of globular cluster evolution within this galaxy. The inner halo includes objects such as NGC 1898 and NGC 2019, which are similar to Milky Way clusters in the inner halo but are more extended and flattened. The oldest clusters have a very extended distribution throughout the LMC, whilst the younger members trace the path of the galaxies bar and highlight recent star-formation regions. Many of the globulars that have been observed in detail show expanded cores, and over 625 of these clusters have been thoroughly catalogued, but estimates suggest there could actually be well over 4,000 members in the LMC alone.

In the SMC, outer halo globulars such as Hodge 11 and NGC 121 are present, and NGC 346, a nebulous cluster in the SMC has an extended HII region and contains at least 33 O type stars. A catalogue of clusters found in an area 2.4 square degrees around the central region of the Small Magallenic Cloud has been compiled and contains data for 238 clusters, with 72 new objects. Photometric data for all clusters was also used to determine their size, member populations, and magnitudes, and complete catalogue cluster members and candidates total over 400 objects.

30 Doradus

30 Doradus is a rich star-formation region, and is the closest known super star cluster region or SSC, situated within the LMC, which is one of the closest galaxies to Earth at a distance of 170,000 light-years. Due to its close proximity, its content can be resolved more easily than more distant galaxies and a wealth of observational data can be obtained in unprecedented detail. 30 Doradus is actually classified as a nebulous cluster, or emission nebula with over 100 million solar masses and is also known as the Tarantula Nebula or NGC 2070. A compact cluster of stars, known as R136, resides at the heart of 30 Doradus, which contains extremely massive and young blue supergiant stars. New technology in the infrared allows dust clouds in the nebula to reveal previously hidden sites of star formation, which show voids that are produced by highly energetic winds originating from the massive stars in the central star cluster. Super star clusters, associated with massive starbursts, are usually associated with dusty and distant galaxies, but 30 Doradus allows astronomers to view these systems close-up. This object has also been dubbed a "super association".

M31

The Andromeda Galaxy, M31, is the largest Local Group Galaxy, and has undergone many surveys and observations to expand the knowledge of its star cluster population.

A recent survey found 43 new globular clusters using x-ray observations based on the examination of x-ray binary stars, which confirmed that the brightest globulars are at the greatest distance from the galactic bulge. The globular clusters in M31 emit high x-rays, are optically brighter, and have a higher metal content than previously observed globulars. Neutron stars are also common in these regions, and there is a possible black hole looming within the galaxy's inner core.

More than half of all the x-ray binaries observed in older galaxies appear within these globular clusters, and it has been suggested that all low-mass binaries were formed in globular clusters, and the non-associated stars were somehow pushed out. One of the recently found globulars is the most distant found so far at 55 kpc from the galaxy itself.

Probably the most famous globular in M31 is Mayal II, also known as G1. Red giant stars have been observed in G1, which are probable members of the cluster, but some could, in fact, be background stars. Most of these giant stars are metal rich so they probably originated from the disk of M31 and not from G1 itself. The formation of G1 is not known precisely, but one possibility is that it was forged from a collision between M31 and a passing dwarf galaxy. In this scenario, the outer extremities of the dwarf galaxy are stripped away, but the nucleus or core survives the interaction and the remnant is the globular cluster we see today. This is a similar scenario to the M54 globular in the Sagittarius Dwarf galaxy discussed earlier in the chapter. The core of the G1 cluster has a comparable composition and orbital motion to typical dwarf galaxies, which gives further support for its formation theory. Although G1 lies at a large distance from M31, many of the other globular clusters in this system are closer to the galactic disk, and run the entire length of the disks central plane, which is evidence that M31 had a large disk in its early evolution. Recently, a suspected intermediate mass black hole hiding within the G1 globular was discovered, which provides more evidence for the link between globulars and dwarf galaxies, as this object is easier to comprehend if it originally occupied the core of a galaxy.

Observational studies show that most of the globular clusters within the vicinity of M31 have similar ages to those within our own Milky Way Galaxy. The globulars within the disk appear to have remained undisturbed from an early age, implying that M31 has not been inflicted with any mergers for some considerable time. M31 also has an elliptical satellite galaxy, M 110 also known as NGC 205, which has 8 catalogued globular clusters. The latest cluster studies on M31, which resulted in the publication of the "Bologna Catalogue" found over 690 new globular clusters and candidates, observed using infrared and optical wavelengths. It is the most in-depth study of extragalactic clusters to date and has catalogued over 1,164 objects, with nearly 600 confirmed clusters and over 600 candidates that require further study.

Fornax Group

Many globular clusters have been detected in the Fornax group of galaxies, and they are mainly old and metal poor. The results of a search for clusters in this vicinity, sampled 11 galaxy members, and detected 237 clusters. Of these, 25% were blue clusters that are probably HII regions, and the remainder of the cluster candidates are massive young and old clusters. A comparison between the red cluster candidates in these galaxies and Milky Way globular clusters shows them having similar luminosity distributions, but the red cluster candidates have larger cores, and are probably globular clusters. There are also at least 468 globulars around NGC 1399, which is the central galaxy of the cluster. A radial velocity study on the group is one of the largest undertaken so far and NGC 1399, the brightest elliptical galaxy, contains clusters travelling from 800 to 2000 km/s. Some of these objects may be unbound clusters or foreground objects that are not related to the group, but red and blue clusters have also been observed in this area, showing variation in cluster ages.

Sculptor Group

There is somewhere in the region of 150 globular clusters and candidates within the Sculptor group of galaxies, and they have a similar age to the older Milky Way globulars, and are metal poor, but have more metal content than typical objects in our Galaxy.

NGC 253, one of the prominent galaxies in this cluster, has an active star-forming region, where at least four very young super star clusters have been observed. This group also provides further evidence that many old globulars were formed in the disk, as opposed to the halo. 17 new globular cluster candidates in the member spiral galaxy NGC 300 have also been located as well as 18 OB associations with more than 30 member stars, many of which are blue supergiants. Over 117 stellar associations have already been catalogued for this galaxy.

M82

In M82, over 24 super star clusters have been observed, but the interstellar dust in this galaxy is probably obscuring many other candidates. M82 also has a range of open and globular clusters within it. The SSCs in M82 contain up to a few million solar masses, an order of magnitude higher than typical globular clusters, but they occupy similar volumes and are, thus, substantially denser than other globulars.

NGC 5128

An analysis of NGC 5128 in Centaurus with "wide-field deep imaging" techniques has been used to examine the globular cluster population in this galaxy, at 2 degrees surrounding its center. The results found less clusters than originally expected, but 211 confirmed, and over 320 possible new candidates were discovered, that were previously not catalogued.

NGC 2158

NGC 2158, The Centaurus A galaxy is relatively nearby, and is therefore a good place to test cluster evolutionary theories. It is a fairly young galaxy with 138 new globulars recently discovered, bringing the known total to 215, one member G169 being particularly bright. This galaxy probably contains two separate populations of globular cluster, an old metal-poor variety, and a newer metal-rich type, which lends credence to the possibility that the galaxy formed from a merger of two galaxies 3–8 billion years ago.

M101

Very recent observations of M101 show a trio of massive young clusters, which may have formed from smaller cluster mergers. This group was detected inside the diffuse nebula NGC 5461 that is associated with the galaxy, 23 million light-years away. Over a million stars have been estimated in the cluster region, which encompasses all three clusters within a diameter of 100 light-years. It is assumed that many of these objects will eventually form globular clusters, and that the entire structure itself will form into a super star cluster within the next few million years.

Summary

Theories of star cluster formation and evolution are only useful until a more probable theory is developed and proves, through speculation and observation, that the old theory is wrong. There does seem to be a certain amount of ambiguity and uncertainty in cluster research at this moment in time, especially concerning super star clusters and young massive clusters, as mentioned earlier in the chapter. However, astronomers are trying to establish very specific scenarios in objects that are thousands of light-years away, which is no mean feat. There are many new ideas on this topic. For instance, it is now believed plausible that the difference between a low mass globular cluster and a massive open cluster is negligible and largely down to time scale. Even the distinctions between open, massive and super clusters is not as apparent on an extragalactic level, so it is possible that eventually, some reclassification may occur. It is feasible that more than half, or possibly all stars form in clusters, for every galaxy, regardless of its location.

With extragalactic clusters, we are dealing with a totally distinct setting to the Milky Way, so many of the existing models and classifications of star clusters simply break down. Scientists do not face the same obstacles in our Galaxy, because there is little star formation occurring – most globulars are old, and most open clusters are young, which keeps everything tidy. In extragalactic astronomy, the classes are not as relevant because we have young globulars, and old open clusters thrown into the mix. However, star cluster formation and evolution is quite possibly the same across the entire extragalactic void, because observations demonstrate that total cluster numbers, massive clusters candidates, and luminosity values are similar across all the galaxies observed to date.

Chapter 8

Cluster Remnants

Only one possible outcome for all star clusters is assured. Regardless of their size, mass or gravitational power, eventually they will all cease to exist. From the smallest loosely bound open clusters, to the mighty densely packed globulars, these objects cannot survive indefinitely and at some point their stellar members will be lost to intergalactic Space. The open clusters are the first star groups that will disperse their component stars into interstellar voids leaving only a few sparse members in the form of open cluster remnants. Even the giant spherical globulars will ultimately be torn apart, leaving only a measly core of stars within a globular cluster remnant. These events will, however, take millions, or even billions of years, but their evolution and eventual termination can be witnessed even now.

Open Cluster Remnants

In the final stages of an open-clusters evolution, the low mass star members begin to lose their constant battle from the continual tugging of gravitational forces. Over time, the stars begin to segregate and eventually this disruption causes the stars to be driven out. A faint core of compact and giant stars is left behind, which becomes an open cluster remnant, or OCR. These objects, also referred to as dissolving open clusters, are the leftovers of these once luminous star groups, and have been discovered and confirmed in various parts of the Galaxy, albeit in small numbers. Initially, the low-mass stars in these clusters are the first to be ejected; leaving a low population of faint, dim stars that can be difficult to discern from background field stars. Several scenarios are responsible for the slow destruction of these clusters; the main suspect is from interactions with other massive stars during the clusters' normal evolution that depletes the clusters' mass. The second culprit is from liaisons with gas clouds within the disk, such as disk "shocks" that continually strip away member stars, caused by tidal forces.

Open cluster remnants are relatively modest in size (3–4 arc min visually) and contain a small number of massive stars, but no low-mass stars, as these have all been dispersed into the galactic disk. The stars typically reside in a fairly compact core, which is made up from many binary stars that were once the members of large old clusters.

Most open clusters have typical life spans of between a 100 million and 600 million years, but it can take up to a billion years for the cluster to be completely dissolved. Larger clusters can survive for longer periods, but only a very small number, about 5%, are older than this figure. These remnants are, therefore, the middle-aged survivors of much larger original clusters that are entering their final stages of life. It is also likely that we are witnessing two separate stages of events within cluster remnants, some are the skeletal remains of a completely defunct cluster, and others are objects still in the process of ongoing disruption.

To reach the status of an open cluster remnant, the group must lose at least two-thirds of its stars, and this disintegration period can begin when the cluster reaches an approximate age of 200 million years. This phase can last for the duration of the cluster's life, up to a billion years, but typically, the time line from initial disintegration to entire dispersion takes about 400 million years. The lowest mass stars in the core of an open cluster remnant normally demonstrate old age, and this is one way to determine the age of the remnant and its group of stars. Remnants do survive for long periods of time in the form of low population clusters before completely dispersing.

It is known that the final size and structure of a cluster remnant is completely dependent on the objects original or initial mass, and its overall distance from the galaxy. Observations for cluster remnants are based on star counts, in relation to known quantities of existing open clusters and field stars, and also using photometry, both in optical and infrared wavelengths. Genuine remnants should demonstrate a noticeable difference in comparison to background star fields.

To date, only a handful of certified open cluster remnants have been confirmed, but there are many candidates, which have some uncertainty, and require further study. (See Possible Open Cluster Remnants below.) Many of the existing remnant studies have found objects that are not true remnants; in fact, several of those observed were not even genuine clusters! Two of the main focus areas are: to determine where dissolving clusters are likely to be found; and to discover whether these remnants can actually be observed.

A recent study of evolved open clusters highlighted NGC 2180 as a possible cluster-remnant candidate. The NGC catalogues this object as an open cluster, in Orion, and the cluster demonstrates signs of star erosion and mass segregation (redistributing stars of different masses and distances) from tidal forces tugging at the cluster. It lies at a distance of approximately 0.1 kpc, so is actually closer to the galactic plane than many normal open clusters. NGC 2180 is believed to be a "halfway-house" object, an old evolved cluster, but not yet a fully blown remnant. With an estimated age of 710 million years, it is relatively young for a remnant, but with a diameter of 9.5 pc and core radius of 0.7 pc it is still relatively large.

NGC 1901 is a confirmed open cluster remnant in Dorado, and is similar in age and structure to the infamous Hyades cluster in Taurus. It is a loosely bound conglomeration of low density stars, at a distance of 0.45 kpc with an approximate age of 600 million years. Its central core of stars lies in the vicinity of the LMC, and 16 of these stars are still on the main sequence, which has enabled astronomers to show the remnant's evolution. Of the 15–16 stars that remain, statistical calculations show this cluster could have originally had over 500 member stars, but this is only an estimate. During the course of these recent observations, it has been suggested that the Hyades is, in fact, an open cluster remnant. The group has lost over 90% of its member stars and fits the bill in terms of criteria, but further studies may be required to fully confirm its status. One of the major hurdles in these investigations is the subtle observational difference between an evolved open cluster and a remnant, which is almost indistinguishable.

When Ruprecht 3, a catalogued open cluster in Canis Major was targeted for cluster remnant study, the spectra of this dense congregation was unusually high in metal content, which suggested Ruprecht 3 was actually a metal-rich globular cluster! With an age of over 1.5 billion years old, and a distance of 0.75 kpc, it is certainly ancient and distant enough, and astronomers even hedged bets that the object could be a globular cluster remnant. Following photometry observations, however, the confusion was probably caused by the clusters' spectrum resembling a spherical structure, due to the lack of any bright, shape-defining stars. It now appears that Ruprecht 3 is more likely to be an intermediate age open cluster remnant.

Two previously unstudied remnant candidates, NGC 7772 and NGC 7036, have been the subject of recent observations, and are both confirmed as genuine open cluster remnants. NGC 7772 was originally catalogued as an open cluster, and new data suggests it is extremely old, at 3–4 billion years, with an immense distance of over 1 kpc (1000 parsecs), and 17 member stars, situated in Pegasus.

NGC 7036 is actually referred to in the NGC catalogue as "non-existent" and this term and other similar descriptions will be covered in the next chapter. As it turns out, NGC 7036 certainly does exist, and is a remnant with 14 members and a visual diameter of 3 arc minutes. With an estimated age of 1.5 billion years, this is another fairly antique cluster, and is even further away than NGC 7772, at 1.5 kpc (1500 parsecs).

Possible Open Cluster Remnants

It may seem bizarre to have a cluster class that only indicates a probable and not a definitive discovery, but this is exactly what the "Possible Open Cluster Remnant" or POCR classification is for, which was recently proposed to differentiate between confirmed and suggested cluster remnant candidates. Observations show that some of these clusters show discrepancies between the size of their major and minor diameters, which may suggest they have an elliptical orbit. These stars also show signs of being flattened towards the direction of the galactic plane.

Several hundred POCRs are believed to exist in the higher plane of the galactic disk, which is also known as the "old disk". However, finding these objects and then trying to distinguish them from other stars or normal open clusters is a difficult and time-consuming task.

Proper motion studies have also been used to probe the inner depths of cluster-remnant candidates, in order to classify their stars as true cluster members. 11 POCRs were recently observed using infrared photometry and comparison data from existing star catalogues, but only one object, ESO 282 SC26 was found to be a real open cluster, 1.46 kpc away. The other 10 objects were classified as non-physically associated stars.

One of the original POCR studies, which observed 34 objects mainly by examining star field populations, showed great promise initially. However, more recent follow-up observations on these objects provide evidence that none of their stars are even related, it is suggested they are simply random groups of non-associated stars. These examples clearly demonstrate the difficulties involved in this field of study, but not all the POCR candidates have been discarded, and there have been some successful findings that will no doubt lead to further discoveries.

Confirmed Possible Open Cluster Remnants include Collinder 21 in Triangulum, with a diameter of 12.7 light-years, and a distance of 7240 light-years. There are also several objects listed as "Turner" after their discoverer, which include Turner 7 in Centauri, Turner 9 and Turner 11, which both reside in Cygnus.

Globular Clusters Remnants

Whilst searching for data on this chapter, I was surprised at how little information is available on the study of Globular Cluster Remnants, but surely they must exist, as all clusters, even globulars have a limited shelf life. The lack of data and observation must be largely due to the fact that many globulars are already up to 13 billion years old, so remnants or extremely old globular "corpses" must be first, few and far between, and second, very faint! Globular Cluster Remnant's are, in reality, similar to open cluster remnants; they are just the remainders of a cluster that has evolved into old age, having suffered countless interactions, star-stripping and tidal forces, to reveal a smaller, faint core of stars. I did find several mentions of these objects in unrelated papers but the data is slim, to say the very least. A globular cluster remnant is certainly depleted in main sequence stars, is low in luminosity and has few or no bright stars. It is likely that these remnants will end their lives as a low-population fossil, eventually being dispersed into collections of double stars or a multiple star system.

Astronomers have, however, made some progress in this area, and have found high densities of stars and clusters that form a ring around the galactic center of a galaxy. One such object was discovered in Canis Major, and has been named the CMa ring. Its component stars have low velocity and contain a high population of RR Lyrae stars (faint variable stars) within the halo region, and its is suggested these could indicate the presence of globular cluster remnants. However, Brian Skiff from the Lowell Observatory is not aware of a single successfully identified globular cluster remnant. Because globular clusters are massive enough to have kept most of their stars since their birth, they still contain enough stars for them to look like globulars, even after the 13 billion or so years since they formed. Skiff notes that there are evidently streams extending from some globulars that seem to have been produced by encounters during galactic plane passages, or perhaps with massive molecular clouds at higher latitude. But even so, Skiff admits that the clusters involved can hardly be classed as "remnants."

One example of this scenario is the globular Palomar 5 which was recently found to have star streams in the form of "tails" both behind and in front of the cluster, that demonstrate the continual removal of member stars during its traverse through the Galaxy.

Summary

Open and possible cluster remnants are a strange breed of object that are not well researched or understood. There have been few studies performed on these often-overlooked classes of object, but they can provide important clues on the evolution and origins of star clusters, and their inevitable demise. Globular cluster remnants have very little or no data available at all, the reasons for which may be explained above. Chances are that cluster remnants will receive closer scrutiny in the future, but much of the current star cluster research is focused on young objects at their birth, or in their prime, rather than ancient "dinosaurs" such as these.

Chapter 9

Misfits and "Nonexistent" Clusters

Modern amateur and professional astronomers are armed with more star cluster data and catalogues than at any other time in the history of observing. These tools allow the detection, observation and collection of material to an unprecedented level. With so much information available, this faces us with a dilemma, as surely some of this data must be prone to inaccuracies or misleading facts. Unfortunately, this proves to be the case, and in this chapter I will try to unravel some of these issues and provide some useful ways to avoid some of the more common pitfalls. There are over 5,000 star cluster and cluster candidates that are readily available to amateur astronomers. Also, this figure incorporates at least 1,600 open clusters or candidates, and 150 globular clusters in our Galaxy alone. As the old computer saying goes, "Garbage Input, Garbage Output," which in this case means that any observations we make today are only as good as the data they were originally based on.

What's in a Name?

Look up a star cluster in any sky atlas or catalogue, or even a magazine and you will no doubt be inundated and quite possibly confused by the different references associated to a single object. Most star clusters, in fact, have at least three or more separate designations that are often added each time an object is reobserved and consequently recatalogued.

The Messier catalogue from the late 18th Century was one of the first official deep sky catalogues, even though it was a list of objects to avoid whilst comet hunting, but many of these objects were reobserved and recatalogued in the NGC (New General Catalogue), by Dreyer, from 1888. Additional catalogues, the IC (Index Catalogue) and IC II followed soon after this, and generally these three catalogues are considered as a whole. Further studies of specific deep-sky phenomena, such as open or globular clusters, have added other definitions to existing objects, and even observations at different wavelengths can result in new classifications for previously, well-known clusters. For example, the open cluster M34 is also catalogued as NGC 2632, and is further known as the Praesepe or Beehive Cluster. Also, the splendid globular cluster M13 is also referred to as NGC 6205, and The Hercules Cluster. Whenever possible, most astronomers use an NGC designation if one is available which standardizes the description of astronomical objects somewhat. Otherwise, a Messier or a more specific catalogue moniker is used. Some observers use their name to classify objects, like Messier, and more recently for example, Alessi, an open cluster designation, and extragalactic clusters also have their own index.

The globular clusters in the Andromeda Galaxy, for example, are called G1, G2, G3 etc, and several extragalactic open clusters are labeled with a C prefix, C1, C2 and so on. When Patrick Caldwell-Moore published the recent (and excellent) Caldwell list of Milky Way deep-sky objects, these were listed as Caldwell 1 to 109, but are abbreviated C1 to C109, which could be confused with the extragalactic clusters already mentioned. This is a typical example of how easily existing objects can be "re-named", often innocently, and can cause potential identification problems. However, none of these object designations are incorrect; they are simply different ways of naming the same object, which can be misleading, but so long as these duplicate names are taken into consideration, they should not cause too many problems.

The NGC/IC Catalogues

Most amateur deep-sky observers begin their journey observing the 109 objects in the Messier catalogue, of which 63 are star clusters, but quickly move on to fainter, more difficult objects in the NGC or IC catalogues. The revised NGC now contains almost 8,200 objects, with over 800 star clusters, asterisms and groups. However, the IC catalogue only contains about 30 faint star clusters. You would assume that these entries give accurate positions, descriptions and observing data on these objects, but to the dismay of many observers and researchers, the catalogues are replete with errors. Duplications, missing objects, "nonexistent" objects that do exist, and listed objects that really do not exist all appear in the revised NGC, but from where do all these contradictions stem? Firstly, the NGC/IC is based on an original set of observations over 115 years old from over 160 different observers. Many objects were simply misidentified, given wrong positions, or even classified incorrectly. The situation was made worse by new catalogues that simply repeated these errors or, in some cases, even substituted a nearby object when the real candidate proved elusive. The NGC was also revised to become the RNGC in the 1970's but this addition actually increased the number of errors. In defence of the NGC/IC, considering the age of these catalogues, and the way they were composed; in the 18th and 19th Centuries from visual inspections, it could be argued that on the whole, the work is pretty accurate. It is still the most widely used deep-sky catalogue today.

NGC/IC Project

Enter the NGC/IC Project, which is a mammoth undertaking by a core of ambitious yet worthy professional and advanced amateur astronomers, who decided to rectify the problematic NGC/IC situation wherever possible. Dr Harold Corwin, a renowned professional astronomer at Caltech, heads this team, which was set up in the mid 1990s in collaboration with advanced amateurs Steve Gottlieb, Wolfgang Steinecke, Malcolm Thomson, Bob Erdman and fellow professional, Dr Brian Skiff of the Lowell Observatory. Between them, the members have spent hundreds, even thousands of hours sifting through data, examining original and historical observations and data, and making comparisons with modern visual and digital studies, such as the Digitized Sky Survey, to remove some of the inconsistencies and inaccurate data. As well as a completely revised and updated NGC catalogue, the team have also solved many solutions to observation problems. The fruits of this labor can be seen on the NGC/IC Project web site, www.ngcic.org

Discoveries and Corrections of the NGC/IC Project

Steve Gottlieb, an expert amateur, searched for the open cluster NGC 6846 in Cygnus, but was unsuccessful with his 17.5-inch telescope under dark skies. With this negative result, Steve sought the help of Connecticut amateur Ernie Ostuno, who had located NGC 6846, exactly 2 degrees north of the RNGC position, using the Palomar Observatory Plates (POSS). By calculating the precession (minor shift of stars over time caused by the gyration of Earth's axis) the original coordinates were updated, and they found the position was an exact match for this small clump of stars. The incorrect position in the RNGC is also found in the Lynga Open Cluster catalogue and is repeated in the Uranometria star atlas.

The description of NGC 2866, an open cluster in Vela is a perfect match of John Herschel's original description of this object, as a small group of stars. Unfortunately, the RNGC description says it is nonexistent, and the Lynga open cluster catalogue and Sky Catalogue 2000 identify the cluster as Pismis 13, a different cluster in the same constellation. The Guide Star Catalogue (GSC) shows the group very prominently and the ESO (European Southern Observatory) Catalogue gives the correct identification, but questions its true identity as a genuine cluster, but NGC 2866 certainly does exist.

The "Nonexistent" Clusters

I first stumbled upon "nonexistent" deep-sky objects in the Saguaro Astronomy Club's database, SAC 7.2, which is another fine catalogue prepared by amateur astronomers. It lists over 10,000 objects of different types and introduced me to this rather bemusing class of object. Within the NGC/IC catalogues there are many objects misidentified or described as "nonexistent" due to a lack of modern observation since their original discovery. The revised NGC or RNGC also cites many star clusters as nonexistent for a variety of reasons. Some of the objects were no longer thought to be genuine star clusters; instead, they were considered just multiple groups containing no physically associated stars. Other clusters could not be detected, photographically or visually, and were not located in existing deep sky survey plates. It now transpires that some of these 'missing' clusters were lost in the myriad of field stars, or "washed out" on photographic plates, so faint that they simply did not reproduce. There are also objects that really are nonexistent but for some reason they are repeatedly misquoted in subsequent catalogues as real objects.

Star clusters in particular have had their fair share of mis-representation over the years. In light of this, the Webb Society, a specialist deep sky club in the UK commissioned a ground breaking monograph "The Nonexistent Star Clusters of the RNGC", written by professional astronomer and prolific amateur observer Brent Archinal of the US Naval Observatory. This work further highlights the problems of inaccurate observational data and presents 229 objects that are proposed as nonexistent, with revised data and discussion where applicable for recovered objects and confirmed clusters. The catalogue includes extensive notes on many objects, with corrected and expanded information along with modern observations of troublesome clusters, complete with in-depth visual descriptions. Archinal confirms at least five duplicate entries, for example NGC 2239 is the same object as nearby NGC 2244, and suggests that 99 out of the original 229, do exist in one form or another, but are not necessarily all clusters. Of these objects, 124 have insufficient data to support or disprove their existence either way. Contributors to this project include Alistair Ling, Harold Corwin and Brian Skiff, and the author encourages observations from external sources to improve the catalogue further.

Modern observations of the cluster candidates were undertaken by the author, and several prominent observers, including those listed above. Using medium-sized telescopes, and data from star atlases, professional and amateur observing catalogues were compared with digitized survey images. From the entire sample of "nonexistent" clusters in the RNGC, amazingly only 6 can definitely be described as nonexistent, and the remaining 124 objects that cannot be verified required further study. However, just because these objects cannot be found does not necessarily mean they do not exist.

Another extremely useful resource for misrepresented clusters is the book "Star Clusters", also by Brent Archinal and Steven Hynes. This is the definitive catalogue and reference manual on star clusters and makes an ideal companion to the book you are holding now. The new edition improves on much of the data in the Webb Societies Monograph on "nonexistent" star clusters and contains over 5,000 clusters and candidates within its hefty tome, that are well beyond the scope of this book.

Examples of "Nonexistent" Star Clusters in RNGC

NGC 886, an open cluster in Cassiopeia, has had a chequered history, and was originally identified as Stock 6, a separate cluster in the Lynga Catalogue, and so the RNGC listed it as nonexistent, even though the original NGC description and coordinates were basically correct. Later observations reinstated NGC 886, but in the 2005 open cluster catalogue (Dias) it has finally been deleted. NGC 886 has been removed because it really is a duplicate of Stock 6, so on this occasion the RNGC was correct. To add further confusion, this object has also been previously misidentified as the galaxy NGC 863.

NGC 1798, is an example of a completely valid open cluster that is portrayed in the RNGC as nonexistent. Brian Skiff, who resolved 15 member stars in a faint cluster, has visually confirmed its position and classification in the constellation Auriga. This object is also listed as Be16 in the Lynga Open Cluster Catalogue, but was published incorrectly in the same "Alias List" as NGC 1789, a totally separate cluster in the LMC, probably due to a typographic error.

Burnham and others have previously listed NGC 2158 in Gemini, as an open cluster, yet the RNGC describes it as a globular cluster. It is actually an old open cluster that can be difficult to resolve, so the RNGC classification is wrong. Also listed as OCL 468, and Mellotte 40, it contains about 900 stars but the brightest is only Magnitude 15, hence the confusion.

The open clusters NGC 6216 and NGC 6222 have caused much confusion over the years, with the RNGC in particular which states that NGC 6216 is duplicate of NGC 6222, and therefore NGC 6216 does not exist. Lynga, on the other hand, only describes NGC 6216 as a cluster with 40 stars, the brightest being magnitude 12. In the Webb Society Monograph, Brent Archinal mentions that a new search should be made to recover NGC 6222, as the original observer Herschel noted two clusters in this area. In 2005, NGC 6222 was deleted from the standard open cluster catalogue (Dias), and is now listed as an asterism, a group of non-associated stars.

As well as citing nonexistent clusters, the RNGC also has some classification anomalies. One such example is NGC 6328, which for some unknown reason was listed as a nonexistent cluster when, in actual fact, it is a galaxy and is defined as ESO 102-G0003.

Finally, NGC 1498 was the only cluster Archinal examined in the RNGC that was probably nonexistent, though even he stated the confirmation as "suspect". It turns out that the RNGC may be correct on this occasion, as there is no "cluster" at the specified location. However, the object in question is probably the triangle of three stars centered about 2 arc minutes west of the NGC position. Assuming the asterism is the object originally observed by Herschel, it is simply 3 stars about 40 arc seconds across, although the faint field stars may have given the impression of more depth to this object.

Missing Clusters – Problems and Solutions

In terms of observing star clusters, these misidentified and nonexistent clusters can cause problems in detecting and collating new observations. If the data is unreliable, you may find that a cluster is not in the correct location or cannot be found at all, but you would not necessarily know if this is due to your own mistake, or an error in the data used to find the object. Similarly, if you want to observe a globular cluster, but instead you find an open cluster in the suggested location, is this a fluke, human error or simply incorrect catalogue information? Worse still is a situation where you completely dismiss observing a cluster because you think, and are reliably informed, that it does not exist, when quite clearly it really is there for all to see! Fortunately, the remedies to these situations are well underway, and the number of missing or misidentified clusters is constantly reduced as new observations and data pours in. There is much that can be done by the observer to improve this situation.

To confirm or disprove the existence of a star cluster, some investigative research is required. The source of the original discovery and the data associated with it are important, and these can be cross-referenced with modern data that is readily available. If the object cannot be visually observed, archived photographic plates or digital sky images can be studied to locate the object's position. Observers should also familiarize themselves with the various star cluster catalogues that are freely available from the Internet, and are constantly revised and updated by their creators. A comprehensive list of catalogues and cluster data resources can be found in Chapter 16.

Cluster Misfit Examples

One famous example of a cluster misfit is the Coathanger cluster, or Brocchi's Cluster in Sagittarius, which had been catalogued for many years as an open cluster. It is not a true cluster, but an asterism, or group of unrelated stars. Another possible mistaken identity is the Palomar globular cluster in Cepheus, Pal 1. It is described as a globular in many catalogues, and several observers have described it as a very small, faint example but evidence suggests it may not even be a globular cluster, but an ancient open cluster. On the other hand, NGC 6540 was originally thought to be an open cluster, but modern observations show that it is, in fact, a globular cluster in Sagittarius.

Another globular cluster, Grindlay 1, catalogued in Scorpius, has recently been completely removed from current catalogues, as apparently, it does not exist; though it is plotted on the Uranometria Star Atlas.

The previously catalogued open cluster M73, also known as NGC 6994 has also been recently demoted, and is now described as just a group of 4 stars, and not an open cluster. M73's status has been challenged many times previously, but recent observations using proper motion studies, suggest the member stars are not related, but even now there is still some doubt regarding the latest data, as some motion studies can be unreliable.

NGC 5385, NGC 2664, and Collinder 21, all previously listed as open clusters, have also been reclassified. All three are now thought to be random alignments of field stars or asterisms.

Photometry and spectroscopy investigation of the open cluster NGC 6738 in Aquila has been performed in order to reveal its true nature. The object contains no defined main sequence stars and proper motion studies show random distances when they should be similar in a genuine cluster. The luminosity of this "cluster" also matches the field stars in the vicinity, so NGC 6738 is definitely not an open cluster, just a bright group of physically unrelated stars. And finally, two other clusters have apparently "swapped" classifications. Whiting 1 was listed as an open cluster but is now catalogued as a globular, and AM-2 (Arp Madore) which was originally defined as a globular has recently been reclassified as an open cluster.

Summary

In astronomy, you should never take anything for granted. Stellar objects can 'change' locations, classification, and even object type within the blink of an eye. As with any science, ideas and theories change, so although we sometimes appear to have inconsistency of data, overall, due to the efforts of many, the information we have to hand is constantly changing, yet improving. In the future, we may come to a point where almost all the astronomical data for the NGC/IC and similar catalogues is near perfect. But we should always leave room for those little anomalies, if nothing else a perfect catalogue would be somewhat boring, and leave little chance for serendipitous discoveries.

Part II

Observing Star Clusters

Chapter 10

Instruments

Naked Eye

When discussing astronomical instruments, it is easy to overlook one of the most impressive optics of all – the naked eye. It is amazing just how much can be seen with the dark-adapted naked eye under the night sky. There are several globular and many open clusters that can be observed without optical aid, including the Pleiades and Hyades open clusters, and the globular cluster Omega Centauri. Stars around Magnitude 6 can be viewed with the naked eye under dark skies, and after about 30 minutes when fully dark adapted, you can expect to see at least another half or full magnitude. When the eye is first presented with a dark environment, we are temporarily blinded. This is because our eyes contain cones which work in daylight, and rods that operate in low light and dark conditions. It takes the rods some time to adjust to the new light conditions, but the longer we stay in darkness, the better the eye's ability to discern faint objects. The human eye is a complex optical instrument that can distinguish many items in minimal light conditions, hampered only by its small "aperture" of around 6mm, which is the average diameter of the eye's pupil. When you compare this with even a small pair of binoculars with an aperture of only 35mm or 50mm, it is astonishing just what can be achieved without optical aid.

Binoculars

However, we must be realistic and without an optical instrument of some form, we are limiting ourselves severely in terms of what could be seen if using modest equipment. It has been said that a decent pair of 8 × 50 or 10 × 50 binoculars will out-perform a cheap 2.5" or 3" telescope, and I would certainly agree with this. Binoculars are inexpensive and very versatile, they have no set-up time whatsoever – just point and look, and are lightweight and easy to use. For astronomy in the suburbs, 10 × 50 are probably the optimum configuration for small binoculars, providing 50mm aperture and 10× magnification. A surprising number of open and globular clusters can be seen in binoculars of this size. In fact, many open clusters are actually better in binoculars than a telescope, due to their large surface diameter. You should not however, expect to be able to resolve globular clusters or pick out faint nebulosity, with an instrument of this aperture.

Small binoculars are also useful for determining a star field when tracking down faint objects, especially if the finder on your telescope is typically small, say 6 × 30.

I often use my 10 × 50s to confirm I am in the correct part of the sky, prior to star hopping with the finder and main telescope. If budget allows, moving up to "astronomical" size binoculars such as the popular 20 × 80 size (80mm aperture, 20× magnification) will show many clusters in greater detail, and several globulars can be resolved into their component stars. However, large binoculars can be quite heavy and cumbersome, so for prolonged use, a tripod or monopod will be an essential item. This does take away some of the "point and look" accessibility of binocular astronomy, but even with tripod-mounted binoculars, there is something attractive and, I suppose, natural about observing with both eyes. A good pair of 20 × 80 Russian binoculars will reveal a vast amount of open and globular clusters and is suited to most amateur's budgets.

Telescopes – Refractors and Reflectors

Even though today's telescopes cover a massive range of models, there are essentially still only two main types of telescope; the refractor, which uses glass optics, and the reflector that utilizes mirror-based optics. There are many combinations and variations on these two concepts but all telescopes are based on these two types. A lifetime could be spent arguing the virtues of the reflector and the refractor, and there are many fans of both types. Fundamentally, they are both excellent though somewhat different in design and specification. Pound for pound spent, the reflector will be cheaper, as mirrors are more cost effective than lenses to manufacture. Also, much larger mirrors can be produced than lenses due to weight and mounting problems. A lens can only be supported in an optical assembly around its edges, whereas a mirror which is much lighter and opaque, can be supported across its entire area. For these reasons, you see very few refractor's above 6", but there are many amateur reflectors at 16", 18" and even 20" aperture!

The question of quality is slightly harder to answer, as it depends on many factors, including materials used in the manufacturing process, and ultimately how much you spend on telescope optics. However, there are some useful basic ground rules. Refractor's come in two main varieties: the low-cost achromat, which uses one or two lenses for each objective made from crown or flint glass; and the more expensive apochromat, or APO that has two or more lenses manufactured from higher quality glass. In a cheap achromat refractor, you will often see blue or red fringes around bright objects, especially on fast telescopes (f8 or lower). This is due to the fact that light travels at slightly different wavelengths for red and blue and these color fringes are a result of this phenomenon, known as chromatic aberration. In the more expensive apochromat's, this problem is overcome by using two or more lenses, made from exotic glass, such as quartz or fluorite. These lenses are mounted in such a way that they compensate for the lightwave errors, and any false color is eliminated. This comes at a cost, though, and a quality 4" apochromat could cost up to £2500. The view through an apochromat, especially fast (f5 or f6) models are unsurpassed, with crystal clear detail and wide field, pin-point star images, but quality always comes at a price.

Refractor's also have another advantage over reflectors, due to their unobstructed primary objective which allows all the light to pass through the to the eyepiece. Unfortunately, all reflectors have a secondary mirror or lens mounted at

Figure 10.1. The authors 4" f5 refractor, a great wide-field telescope.

the objective end of the telescope – this reduces the amount of light that enters the scope, typically between 25 and 35%, which results in some loss of detail, albeit fairly minor in terms of deep-sky observation.

The reflector is probably the most popular telescope in deep-sky observation due to the relatively low cost per inch of aperture and the quality of the instruments available. Reflectors do not suffer from chromatic aberration because a mirror transmits light at all wavelengths equally – so no color fringing can occur. However, reflectors suffer from different problems, the main ones being collimation and spherical aberration. The most popular reflectors on the market, the Newtonian's, are especially prone to collimation problems which lead to unsharp star images and generally poor views from the telescope. Collimation basically ensures that the primary and secondary mirrors are aligned correctly, which can be performed visually or using a collimation tool and is very quick and easy to do. Under normal usage, the collimation should stay in place for months or even years, but any bumps or knocks, when moving or transporting a reflector, can cause the telescope to be misaligned. Another affliction of the reflector is spherical aberration, which comes in many flavours but is harder to overcome. Problems such as under and over correction or pincushioning are errors in mirror manufacture, but are not really worth worrying about. It is important, however, to keep a Newtonian's optical tube capped when not in use because of the open tube design, dust and debris will eventually tarnish the aluminium coating if the mirror is not properly protected.

Catadioptrics

The word catadioptric describes a hybrid telescope that combines the use of lenses and mirrors, theoretically providing the best of both worlds. The result is a compact, lightweight telescope that overcomes some of the problems of lens or mirror only designs. The professional Schmidt camera telescopes were the forefather of perhaps one of the most popular telescope systems available today, the

Figure 10.2. The author and his 8" f6 Newtonian © Janette Morgan-Allison.

Schmidt Cassegrain telescope or SCT. Whilst the original Schmidt telescopes were around f4, giving wide angle flat fields to photograph the sky, today's SCTs are normally f10 but still have a flat field of view. The Schmidt Cassegrain uses a mirror for the primary objective, but the secondary mirror is mounted to the reverse of a thin flat Schmidt lens, which corrects the light before it hits the primary mirror. The light path enters the telescope, hits the primary mirror, then bounces back to the secondary mirror, where it is then sent back to the primary, through a hole in the mirror, and onto the eyepiece. This extended light path results in a very long focal length in a compact optical tube; for example, a focal length of 1250mm can be squashed into a tube only 350mm in length, because the light travels up and down the tube three times before entering the eye. As with the Newtonian design, the SCT suffers slightly from the central obstruction caused by the secondary mirror, but as previously mentioned, the resulting loss in contrast and detail is negligible for star cluster observation. SCT telescopes were introduced in the late 1970s for amateur use, but are now one of the most popular designs combining compact size with high quality.

Another popular catadioptric configuration is the Maksutov Cassegrain that is similar to the SCT but has a thick curved meniscus lens at the front of the telescope, as opposed to the Schmidt corrector lens. The Maksutov has a small silvered spot on the back of the meniscus lens, which sends the light back through a hole in the primary mirror for focusing. Due to the spherical nature of the corrector lens, Maksutov's are generally considered to be "slow" systems; so focal ratios of f13 or f15 are common. The Maksutov or "Mak" is popular with planetary observers because larger f numbers mean longer focal lengths which, in turn, result in higher magnifications. This also means that higher f numbers provide a smaller field of view. Some observers may consider that Maksutov's or "Maks" are not suitable for observing clusters, but this is not strictly true. They will provide a smaller field of view than many telescopes, so large open clusters cannot be viewed in their entirety, but the extra magnification will be extremely useful when trying to resolve globular clusters or very faint open clusters.

As with all optical instruments – there is no one perfect telescope that can perform well on every astronomical object. Unless you can afford to buy several different telescopes to meet different objectives, it is more likely you will buy an "all rounder" and then compromise, by making the best of the equipment you have, and by trying different eyepiece and barlow lens combinations.

One of the great advantages of the SCT and the Maksutov is that the focus and eyepiece is at the rear end of the telescope, unlike Newtonian's where focus is at the top of the optical tube. This makes observing much easier as the eyepiece remains at a reasonably low level, also attaching accessories such as filters or CCD cameras is much easier. You will also find that the actual focusing mechanism of a Schmidt or Maksutov is very smooth, due to the micro-focusing engineering employed, which actually moves the primary mirror to achieve focus. Instead of turning the focuser frantically, then backtracking as you often do with refractors and Newtonian's, the SCT and Mak designs provide a very positive and precise focus method. As both systems are closed tubes, this also keeps all the optical surfaces clean, dry and free from dust, which keeps maintenance down to an absolute minimum.

There are also several less popular variations of catadioptrics available such as the Newtonian Cassegrain, the Klevsov Cassegrain and the Ritchey Chrétien, which also use mirror and lens combinations in a slightly different design, but they essentially provide similar performance.

Rich Field Telescopes

In recent years, the Rich Field Telescope or RFT has become a popular choice for the deep-sky observer and imager, due mainly to the wide field of view available. This type of telescope is a lens-based refractor, and as mentioned earlier (see Refractor's) the RFT comes in two types – the expensive apochromat, and the low-cost achromat. All rich-field telescopes have low focal ratios, in the region of f5 to f7 and, therefore, have compact tube lengths, which make them ideal for travel in this country or even abroad, as hand luggage. Many models on the market can use both 1¼" or 2" wide angle eyepieces, and combined with the low focal ratio you can quite easily achieve wide field views of 3° or more. This type of rich field astronomy is ideal for large clusters and other extended deep-sky objects, and is perfect for astro-photography or CCD imaging. Several prominent companies renowned for world class optics have begun to specialize in these rich field models which come in a range of aperture sizes from around 3" to 6". With their unobstructed optics and exotic glass, these high-end instruments provide exceptional views, but even the cheaper "clone" imports from the Far East provide perfectly acceptable results.

Light Buckets

As deep-sky work is all about finding dim objects and observing faint detail, the "light bucket" is certainly the cure for aperture fever. Ten or fifteen years ago, an 8" telescope was considered to be fairly large but today there is another very different trend in amateur telescopes. Where the RFT is portable, convenient, and emphasizes quality, the "light bucket" is all about size, light gathering power and sheer mechanical volume.

I'm not sure where the term "light bucket" originated, but essentially we are talking about large telescopes, generally Newtonian in design, in the 12" and larger aperture size bracket. There is a market, albeit small, for large Schmidt telescopes but few can afford them – this is a hobby after all! Most of the large Newtonian's are 12- to 20", but there are some amateurs with scopes of 24" and even larger. Telescopes such as these can be equatorially mounted (see section on mounts) but for a scope of this size, this could prove difficult, not to mention the size and weight implications. A far easier option is the Dobsonian mount, which is a simple up/down, left/right system similar to the Alt-Az system explained later in this chapter. The telescope tube is mounted in a cradle that swings up and down and, this tube in turn, rests on a rotating platform, providing left to right movement. You can buy off-the-shelf Newtonian / Dobsonian configurations at sizes up to 12", but there are several companies who will custom build these systems to suit the users requirements and budget, to pretty much any size – within reason! Obviously, these systems need serious consideration in regard to siting and usage, but a scope in the 14" to 18" aperture region will provide some stunning detail. Size comes at a cost though, and often with a large Newtonian you will need to stand on a box, or even use a stepladder just to reach the eyepiece! You have been warned!

Aperture Size

Amateur telescopes are now available in a vast range of aperture sizes, from the humble 2.5" (60mm) refractor, to the monster Dobsonians at 20" (500mm) and larger. With such a wide selection in the market, what size is best? Without question, the larger the optics, the brighter and more detailed the visual image will be, and the more magnification can be used. As you increase aperture, however, you also increase the amount of stray light, light pollution and atmospheric effects, so there is a cut-off point when larger telescopes begin to trade-off against smaller models. A 2.5" or 3.5" scope, on the other hand, is a bit too small for deep-sky work, so an ideal starting aperture is about 4" or 105mm. This size will enable stars up to magnitude 12 to be observed, and can resolve many star clusters. One step up the aperture ladder is the medium-sized scope in the 6" to 8" region (150 to 200mm), these models are larger but still fairly compact, resolving stars to magnitude 13. The 200mm or 8" Newtonian or SCT has been considered the workhorse of amateur astronomy for several years, as it provides a decent aperture in a modest frame. Larger telescopes in the 10" to 12" (250 to 300mm) region certainly open up the sky even more, to magnitude 14, but become heavy and cumbersome, suited more to permanent locations and setups. Finally, even larger telescopes, from 14" up to 20" (350mm to 500mm) can be considered research instruments, especially at the 500mm or half metre range. The cost of purchase, installation and maintenance of this type of scope makes it prohibitive for almost all amateurs, but at this aperture you can locate faint objects smaller scopes just cannot detect, down to magnitude 15 for visual observation. A very large telescope can also deliver subtle detail that small optics cannot resolve, and if you ever get the change to look at a globular cluster through an 18" or 20" telescope, you will not be disappointed – the myriad of stars will astound you.

Telescope Mounts

As we are all observing from a moving platform, the Earth, tracking objects in the sky can be tiresome, which is why many observers use an Equatorial mount. This ingenious device is basically set up to your latitude and aligned perpendicular to the Pole Star, so the mount can then compensate for the Earth's rotation. Once an object is located, the Declination (up and down) is constant so only the RA (Right Ascension) is moved to track the object in the sky. With a manual drive, you simply turn the RA knob forwards (or backwards) to keep the object centered in the eyepiece. This works very well for visual observation and as you only have to alter one axis to keep tracking it is easy to achieve, however to track objects for photography or CCD the polar alignment must be precise (see Chapter 13 on Locating Objects).

Bear in mind that the larger the magnification, the faster objects will move across the field of view, so using a power of 250×, the object you are viewing can "disappear" in a matter of seconds. This is where an RA clock drive can be of great use, as an electronically controlled motor takes care of tracking for you, by precisely moving the RA axis by the correct amount. Both telescope axes also incorporate a dial, which contain the Setting Circles that provide numerical feedback of your co-ordinates in the sky. Right Ascension drives and Setting Circles will be discussed in detail in a later chapter.

All Equatorial mounts have the telescope tube offset, to point towards the Pole Star, and for this reason the scope must be counterbalanced with weights, which are normally supplied. This increases the weight of the overall setup, but because the tube is perfectly balanced, you can move it with the touch of a finger, even on very large scopes. The tube can be moved with or without the axis locks in place – so it all makes for a very smooth and light operation. There are, however, a few disadvantages of Equatorial mountings. Moving the scope near or around the Zenith or the Pole can be difficult, and viewing very low or very high objects can put the eyepiece (especially on a Newtonian) in some strange positions!

If the Equatorial mount is too complex, or perhaps even overkill for your own observing experience, perhaps an Altitude-Azimuth or Alt-Az mount would suit you better. An Alt-Az mount is a much simpler system that allows the telescope to be pointed anywhere in the sky without the need for alignment or tracking. The telescope basically just moves up and down, or left and right, and to follow objects in the sky you must nudge the scope in both orientations. Most mounts of this type have locks on both axis to prevent the tube from slipping, and some have clutch type mechanisms that stay in place unless the user exerts some force to change the orientation. There is nothing complicated about the Alt-Az mount, and it is certainly easy to use, but it cannot be electronically driven to track objects (except on specially designed GOTO Telescopes; see later section). Several Alt-Az mounts do, however, have slow-motion knobs that allow precise movement of both axes, but there are no setting circles or other indicators to tell you where you are pointing. The simplicity of the Alt-Az mount is also used in the popular Dobsonian telescopes that provide a point and look approach without the frills of complexity and automation. Although very recently, even the humble Dobsonian has been upgraded, with one company that has launched a range of computerized "push-to" versions that provide digital readouts on your position in the sky.

What Should I Buy?

There is, unfortunately, no "best" telescope, and no "best" mount! Each user has his or her own requirements, expectations and budget, so it is difficult to recommend any single telescope outright. One of the best ways to hone through the options is to try out as many telescopes as you can at an open night or observing session at your local astronomy club or society. The best telescope for you, is the one you will actually use! A large 12" telescope might give you excellent views, but if it is stuck in a garage because it takes so long to set up - you actually get no views! Similarly, a small cheap telescope may be portable, but if the views are terrible, it is equally useless. Think along these lines before you part with your hard-earned money.

I once heard a deep-sky observer say there is no substitute for aperture. Personally, I think there is no substitute for hours spent at the eyepiece. It does not really matter what type or size of telescope you use, the main thing is to get out and observe, practice your techniques, push your observing skills to the limit and, most importantly, enjoy the experience.

Chapter 11

Equipment and Accessories

At this current moment in time there is wealth of accessories and equipment available to the modern observational astronomer. With such a vast array to aid and improve our viewing enjoyment and increase our ability to observe a greater range of objects, the choice can be confusing. In this chapter, several examples of the kit currently available have been scrutinized, some which are indispensable such as eyepieces and diagonals, and other items which are certainly useful, but not crucial to a successful observing session.

Eyepieces

Probably the most important accessory, after the telescope and mount, is the eyepiece and this is an area where spending as much as you can realistically afford will pay off. A decent quality telescope is fairly useless if you use a cheap or inefficient eyepiece to view objects with it, and you will not get the best out of your telescope.

Unfortunately, typical deep-sky telescopes, with lower focal ratios (f no.) and wider fields of view are more susceptible to poor quality eyepieces than longer focal telescopes used by lunar and planetary observers, which can result in image and color distortion.

Today's market for eyepieces, or oculars, is wide and varied, and there are models to suit all budgets and requirements, and several telescopes are often sold with one or two eyepieces as part of the package. A popular entry-level eyepiece is the Kellner that has a fairly wide, flat field and is also known as a Ramsden. They are relatively cheap and especially useful for longer focal ratios, but high magnifications can lead to some distortion. The Plossl eyepiece, which is available from many different manufacturers, is possibly the most widely used lens by amateurs, and offers medium to high quality performance at reasonable cost. Plossl's are available in many different sizes, from 2 to 3mm focal lengths up to 40mm ultra wide-angle models.

Orthoscopic eyepieces were popular many years ago, but have made a comeback recently and provide high quality and high magnification with little distortion and good eye relief, although some models can be quite expensive.

At the premium end of the market, the Naglers, Panoptics and Pentaxes dominate. With their multi-element design they provide tack sharp wide and ultra wide field angles with virtually no image distortion, and can provide good magnification. However, they are expensive, fairly heavy and quite large compared to "normal" eyepieces of the same focal length. For most observers, a selection of two or three eyepieces and a barlow lens combination (discussed later) will provide ample views, giving low power wide-fields, medium power for general observing and high magnification to resolve detail.

There are two main criteria that define an eyepiece, focal length and field-of-view. Focal length determines the magnification of a lens, but this also depends on the focal length of the telescope. Field-of-view (FOV), as the term suggests, is the amount of sky visible through the eyepiece, and is normally expressed in degrees, generally between 25 and 80°.

A 10mm ocular gives twice as much magnification as a 20mm, but a 40mm eyepiece provides half the magnification of a 20mm. However, the 40mm delivers the widest field-of-view, and the 10mm with the highest magnification, actually has the smallest field-of-view.

Most telescopes have a theoretical magnification limit, which is approximately twice the aperture in mm, so a 150mm telescope can produce a realistic magnification of 300 times. This is, however, only a guide; high-quality telescopes and lens combinations under pristine skies can go further, but under normal circumstance these figures are fairly accurate. The term "eye relief" is also often expressed by lens manufacturers which refers to the distance between the surface of the eyepiece and the point at which the image is formed, called the "exit pupil." Larger eye relief makes for more comfortable viewing, especially for those who wear spectacles.

As mentioned earlier, magnification also depends on the focal length of the telescope.

To calculate the magnification for a given telescope and eyepiece, use the following:

$$\text{Magnification} = \text{Telescope Focal Length} \div \text{Eyepiece Focal Length}$$

For example, 1200mm telescope ÷ 20mm eyepiece = 60×.

Wide-field eyepieces are also readily available that combat the problem of small fields of view, by utilizing multi-element lenses they achieve wide fields with relatively high magnifications and virtually no distortion. Available as WA (Wide Angle) and UWA (Ultra Wide-Angle) versions, these oculars can be very useful when observing large, scattered open clusters and associations.

There are a surprising number of low cost telescopes that now feature 2" focusers that were once the domain of premium models. Most of these also accept the standard 1¼" eyepieces, so you can mix and match. To compliment this a good choice of imported 2" eyepieces is now available that also provides acceptable wide-angle views without the price tags of the high-end models.

Barlow Lens

One simple way to effectively increase your eyepiece range is to invest in a Barlow lens, which simply fits in between your focuser and the existing lens to increase the eyepiece magnification. Using this device doubles your ocular arsenal, so if you have a 10mm, 15mm and 40mm oculars, in theory with a Barlow you can also achieve focal lengths of 5mm, 7.5mm and a 20mm. There can, however, be some duplication. If you own 10mm and 20mm original lenses, using the 20mm with a 2× barlow provides a second 10mm lens, so if you plan to use a barlow, try to invest in lenses that give you "new" focal lengths.

Figure 11.1. A selection of low, medium and high power eyepieces.

Barlows are usually available with 2×, or 2.5× magnification, but 3× or even 4× models can be purchased, though I would be inclined to stick with the lower specification. This setup also allows high powers to be achieved without resorting to small focal lengths and, subsequently, narrow fields of view.

A few years ago, many of these devices were of inferior quality and design, but there are some good quality versions available in both achromatic and apochromat designs depending on your budget. But they cannot match the quality of an original eyepiece of a pre-determined focal length, but are a good enough second best.

Diagonals

A star diagonal is pretty much standard equipment on most refractors and catadioptric telescopes, and they are used mainly to redirect the light path and therefore the position of the eyepiece to a comfortable position for the observer.

The standard size is 1¼", but 2" versions are becoming more popular, allowing the growing range of 2" eyepieces to be fully supported. A diagonal is typically a 90° prism that reorientates the viewing angle at 90° to the telescope, rather than leaving the eyepiece in the same plane as the telescope tube, which is impractical for most observing positions. Several types are available, which include mirror versions, which have an aluminized-reflecting mirror, and prism versions, which use a lens but are more expensive. Both versions result in a slight but negligible loss of light, in comparison to "straight-through" viewing; however, high quality "dielectric" coated models are also on the market which promise up to 99% reflectivity, which results in almost no loss of detail of the light path through the diagonal. A 45°prism is also available, but these are generally suited to terrestrial observing, though some amateurs prefer them for "seated" observing.

Most telescopes deliver an inverted or mirror image of the original object, but when viewing astronomical objects, especially clusters, which are fairly, nondescript, this does not affect the subject matter. However, introducing a star diagonal will alter this optical configuration inverting the image left to right, but again, this not detrimental, just something to be aware of.

Deep-sky Filters

For many observers, the introduction of deep-sky filters made the difference between detecting a diffuse object, and not seeing it at all. Often referred to as nebula filters, these devices have performed wonders on many emission and planetary nebulas, especially from urban and suburban locations. For the avid observer of star clusters, however, these nebula or narrowband filters have little or no effect as they are designed to allow the penetration of particular wavelengths pertinent to gaseous clouds and nebula.

All is not lost though, and there are filters available that can improve the view of open and globular clusters or associations, particularly if light pollution is a problem in your observing locale. Some Broadband filters, also known as Light Pollution Reduction (LPR) or "skyglow" filters can improve the detection of star clusters, but will not improve the cluster itself; in fact, the luminosity of the stars is actually dimmed by a magnitude or so. What these filters can do is enhance the contrast between the background sky and the cluster itself by blocking some of the sodium and mercury emissions from street lamps and other atmospheric light pollution, such as "skyglow". To be honest, the effect is subtle; so in our case, filters are not really a great solution and my advice, if possible, is to try before you buy.

Finders

Deep-sky observers, in particular, star cluster fans sometimes have the need for dual functionality in their observing equipment. Our target objects are large and scattered such as the open clusters and associations, but are also compact and diffuse, like the many globular clusters. For this reason, you may actually prefer to use two separate finders that perform different roles. The optical finder, or finder scope is still popular, and is in effect a miniature telescopes with a low magnification and wide field, typically 6 × 30 or 10 × 50, that allows the main telescope to be oriented correctly. These units are especially useful for defining particular stars within fields that you cannot see with the naked for "star hopping" and navigation purposes. They are not ideal, however, for obtaining a general location, or for aligning the telescope with bright naked eye stars. Most models have a "crosshair" that aids in centring objects, and some observers prefer right angle models that provide more comfortable viewing, as you are looking down, instead of along the tube, but this can be annoying as you are not facing the sky when using the finder. Alternatively, "straight through" finder scopes require some "neck

Figure 11.2. An optical finder.

craning" when swapping between the finder and eyepieces, so both types have strengths and weaknesses.

Unit power finders, such as the Telrad and Rigel's Quick Finder provide better functionality in this area and project a circular illuminated reticle onto the sky which aids navigation, without magnifying the sky above you. They are essentially 1× finders, and provide a non-magnified wide field of view. A more recent addition is the Red Dot or LED finder, which simply casts a bright red dot onto the sky, allowing a wide field to be viewed when aligning the telescope. The beauty of these non optical systems is that they can be glanced at without removing your face from the eyepiece; in other words, you can keep your eye on the finder, and the real sky at the same time, which makes navigating around much more intuitive. Both types of "projection" finder are battery operated and can be mounted to a variety of telescope tubes, though the Telrad is quite hefty! Bright objects work better with illuminated finders, and fainter objects are generally more suited to optical finder scopes. Personally, I think both optical and unit power or red dot finders work well together, and compliment each other, so considering their relatively low cost you can easily mount both onto your telescope tube.

Focal Reducers and Field Flatteners

When observing some of the larger clusters and associations a wide field of view is essential, especially if astrophotography or CCD imaging is to be undertaken. However, many of the most popular telescope designs such as the Schmidt

Figure 11.3. A unit power finder.

Cassegrain (SCT) have relatively long focal lengths in the region of f10. Other catadioptric telescopes, and some of the large tube Newtonian's suffer from the same problem. You can, however, obtain a focal reducer, which fits onto the back of the SCT, to which you attach your eyepieces as normal. The focal reducer uses a clever mix of lens elements to effectively reduce the focal length of the telescope to the region of f 6.3 or f 3.2, depending on which model you buy. These devices are not cheap, but give your telescope dual focal ratios – two telescopes for the price of one! Field flatteners and Coma Correctors perform a similar task but instead of reducing the focal length, they improve any distortions in the telescopes image and go some way to providing edge to edge flatness along the field of view. This reduces any star "stretching" that can occur near the edges of the view, especially on large Newtonians with fast focal ratios of f5 and lower.

Focusers

Off-the-shelf telescopes are normally supplied with the focus unit when shipped, and often you have no choice as to which version is attached, but some manufacturers do give you an option. The Rack & Pinion focuser is fairly standard across the range for refractors and reflecting telescopes, and is simply a geared wheel on the focuser knob that engages with a flat-toothed plinth on the focuser's draw tube. This setup works fairly well but can be a little crude, making fine adjustments cumbersome, and some observers complain about "backlash" and difficulty in obtaining a spot-on focus. Crayford focusers are similar in design to rack and pinion models, they both have a twin-knobbed shaft that moves the focuser tube, but the Crayford design has no gears or teeth. It works using friction from a smooth shaft that presses on a smooth plinth; so the movement is very silky and vibration free, allowing fine adjustments to be made with ease. A third type of focuser, which is radically different to the first two, is the Helical focuser which normally encompasses a draw tube for rough focus, and a collar or ring that you twist to perform fine focus adjustments in a similar fashion to a binocular. An ultra fine focus can be achieved with a helical system, but some observers are not taken with the design, especially for use in the dark, although the system is much more compact than other types. For users of SCT and Mak-Cass telescopes, focusing options are more limited, as these types move the primary mirror, not the eyepiece, but both systems feature micro-focus technology anyway, so focusing is not normally a major issue.

Binoviewers

Binoviewers have been available for several years, but recently new products on the market have reduced the cost of this accessory considerably. The binoviewer allows the user to utilize both eyes whilst observing, which is not only more natural, but also provides a more comfortable viewing experience. Star clusters take on an almost three-dimensional appearance, and some observers have noted that resolution and magnitude limits are improved. The downside is that some units are quite heavy, so your focuser must be able to withstand this weight, and you also require two eyepieces for every focal length of your choice. Some care should also be taken when purchasing eyepiece and binoviewer combinations so as to ensure proper focus.

Anti-dewing Devices

Certain telescopes, especially SCT and Maksutov models are especially susceptible to dewing, although this can occur with refractors and to some extent, even reflectors. Cold clear nights can bring on this dew, especially in the early hours, and it is undesirable, spoiling the views and even damaging precious optics. The simplest correction is to buy or make a "dew shield" which is basically a tube that extends beyond the primary lens or corrector. This prevents the dew from building up, and some refractors have built in shields for this very reason. If you experience more severe dewing, there are several anti-dew heaters available that can prevent build up. Most of these devices are based on a metal strip or cord that encircles the main optic, applying slight warmth that can eliminate dewing problems completely.

Chapter 12

Observation Planning and Resources

Depending on where you live, especially if this is an urban or suburban location, clear nights are becoming increasingly rare, so it is important to plan ahead for these occasions to ensure you can observe at a moment's notice. The three key points for planning an observation session are preparation, preparation, preparation, which cannot be stressed too highly! Some observers take considerable time to set up their telescope, and only then decide what targets are available for viewing, and by the time this has been determined, more often than not, the clouds have conveniently rolled in! By organizing an observing session beforehand, the exercise itself will be much more rewarding and enjoyable, and the fruits of your labor will reflect this planning.

Planning an Observing Session

An organized observer will undoubtedly not only see more objects, but will have the time so really scrutinize these clusters without worrying about what else is available for observation. It is not really worth considering observing without planning somewhat in advance, unless you have limited time and just require a few quick peeks at some of the better known objects. One of the best ideas is to create a target list or "hit list" of clusters that you would like to observe, which can be based on the many existing specific catalogues, such as the Messier or Caldwell list, or cherrypicked from other sources.

Get into the habit of noting objects of interest that are often listed in the "what to see" columns of monthly astronomy magazines and in the articles that often appear on specific astronomical objects. These can be added to your wish list, which can then be given some order or priority. Personally, I prefer to list my target observations in order of brightness or luminosity, with the brighter objects first, so that dark adaptation has had some time to set in before trying for fainter, more elusive objects. You can also scale these clusters in order of difficulty, that can help to define which objects you attempt in relation to how good the skies are. Obviously, bright open clusters can be seen in moderate conditions, but fainter more diffuse globular clusters require decent skies to be observed with any clarity. It can also be useful to compile your target list in order of constellation so that you can observe a handful of clusters in the same region of sky at the same time. Better still, the objects can be listed in order of Right Ascension, so that as the sky moves overhead, new objects are repeatedly brought into view without the need for constantly panning the night sky.

If possible, try to observe objects that are as high in the sky as possible, far from the horizon, so as to avoid sky glow and the haze or mist, which hugs the lower regions of the sky. An exceptionally clear, transparent sky is not necessarily the best for observing, as under such conditions the air is quite turbulent, causing stars to take on a shimmering or "boiling" appearance. When observing clusters, a very transparent sky is not a necessity, as many of these targets are not diffused, and so a little haze should not be too detrimental to the views that can be obtained.

It is best to contemplate your observing list and planning stages when the weather is cloudy, or during the daytime, so as not to waste those precious clear nights. I would not advise doing any organizational work if the conditions are good for observing – your time is much better spent outside, doing some real astronomy!

Sifting through observing lists and catalogues may seem daunting, and from an active observer's point of view, also a little tedious, especially if your real thrill comes from being under the stars with a telescope at your side. But it is definitely worth spending the time on these planning stages, and will actually improve your observing enjoyment and your skill levels, as you increase the difficulty of the target list. It is also rewarding to tick off these objects, which gives a sense of achievement and provides a memory aid for writing up these observations in detail at a later date.

Observing from Home

For much of the time, most amateurs will probably be observing from their backyard or garden, and although this does not present the most favourable viewing conditions, it is certainly more convenient than visiting an external dark sky location. Even when observing from home, a little preparation can go a long way to a successful observing session. One thing to consider is your local horizon, which is often marred by buildings, trees and other obstacles that often completely block the very object you wish to observe, so take this into account when planning an evening's observation. Localized light sources can also play havoc when observing faint objects, so turn off as many external lights as possible and disable any security lights if this function is available. If you are trailing any equipment leads into your home, ensure these are safely secured to prevent tripping over them, or install some of the small red LED lights that are available so that you can see where they are positioned. Most electrical power sources on modern telescopes and mounts are designed for outside use and are protected accordingly, but it is wise to check these periodically. If you are using a laptop PC externally, it may be best to use batteries rather than a mains supply. Fortunately, most digital cameras, webcams and CCD units can also be run on batteries or directly from a PC or laptop which is very convenient. Several manufacturers also now supply tailor-made portable battery packs than can safely power all your equipment from one source.

An excellent alternative to continually setting up and dismantling your equipment before and after every observing session is to build or purchase a permanent observatory of some description. Although you can quite easily acquire a fully computerized observatory dome (assuming you can afford one!) that would not look out of place atop of Mauna Kea, there are several easier and cheaper alternatives. The runoff shed provides an excellent shelter for your telescope and yourself, and can be custom built from "breeze block" or wood, or adapted from an existing garden shed, provided it is large enough to accommodate your telescope. The roof of this type of observatory is detachable from the main body of the building and simply pushes to one side on rollers, with the roof section supported by several pillars. They can be built for a relatively small outlay, and provide the opportunity to have all your equipment setup and aligned on a permanent basis. The telescope and mount can even be concreted onto a pier, which makes it ultra stable. And by running in electricity, all your equipment, computers and accessories can be powered easily – even a red / white light combination can be installed. The advantages of this setup are enormous; simply roll back the roof and you can be observing in a matter of seconds. However, there are one or two drawbacks to consider. Security is a major consideration, as garden sheds in particular are notoriously easy to break into and appear to be a popular target for thieves, and with all your precious equipment outside under one roof, a potential thief could have free run – so prepare for this unfortunate possibility. Another potential problem concerns the siting or location of the observatory, which requires some careful consideration. A permanent telescope cannot be moved which could restrict parts of the sky that could be accessed with a portable model, so ensure the location gives good access to a decent portion of un-restricted sky, preferably favouring an East to West inclination as the stars move in this direction.

Observing Away

A field trip to one of the many, still relatively, dark sites can be a rewarding and enjoyable event but requires some preparation in advance. The telescope and equipment should be thoroughly checked prior to setting off, to ensure everything is in working order, and nothing is absent. Lenses, filters, and other ancillaries are best packed in a tool box or case to keep everything together, and you may want to consider purchasing or adapting some form of case for the telescope itself. Even covering the tube in bubble wrap or similar material is better than nothing, and will go some way to prevent damage or mis-collimation of the optics. Any equipment leads or hand controls, drives and such like can also be boxed or bagged up for convenience. Spare batteries and bulbs for equipment or red light torches are essential, especially if you use a GOTO telescope or have battery operated RA and Dec drives on your mount. A basic tool kit is also a useful, but often forgotten item that can bring life to defunct equipment, and prevent an impending disaster, that could easily be avoided – "If only I had brought a screwdriver" springs to mind. You should also consider taking plenty of warm clothing, with layering of clothes being very important on cold nights. Hats and scarves are useful, and finger-less gloves or mittens allow manipulation of controls whilst still keeping your hands relatively warm. Multiple layers of garments, such as t-shirts, sweatshirts and even thermal underwear can make a significant improvement to your comfort levels.

There are still quite a few dark sites available in many parts of the UK and USA, though even these are becoming scarce with light pollution creeping into even the darkest skies available. Some of these sites are recognized by local and national astronomy societies and have regular meetings or star parties that are open to all amateur astronomers regardless of membership of their club. Other dark sky sites are owned or operated by the societies and require you to be a member to gain access to these facilities. But there are also lots of rural locations, especially in the countryside or in remotely populated areas that have low levels of light pollution and make excellent observing sites. The use of these, however, may be restricted and you should exercise caution if you choose to make use of these areas, especially if you are a lone observer. Standing in a field, in the middle of the night, in an isolated location is not necessarily the safest place to be! Then again, there are unlikely to be many people in the vicinity in the early hours of the morning! It may also prove pertinent to ensure you have permission to observe at your intended location especially if it is remote, and keeping in regular contact with someone at home, via a mobile phone gives some peace of mind regarding your personal safety.

A final duo of useful items for field trips is simply a chair and table – one for you to sit on, and one to keep all your accessories together, off the ground, clean and dry.

Why limit yourself to your own locations mediocre skies, when you can travel the world and sample the celestial delights of other countries? Many seasoned observers are also seasoned travellers, and if you have the right type of telescope, there is no reason why you can't take it with you. Small portable rich-field retractors from 2.5" up to about 4", and some of the smaller SCT or Mak-Cass telescopes make ideal travel telescopes, and most of these models can be carried on a plane as hand luggage with your accessories. Larger items such as tripods or mounts can travel in the cargo hold as they are fairly robust, but the sensitive optics travel with you and should stay in perfect condition. This is a great way to observe star clusters that are beyond your horizon, and it makes a rewarding change to view stars and constellations that are totally different to those back at home. Most manufacturers sell "airport approved" cases and bags that can protect these telescopes without any trouble; however, do not even consider taking a telescope abroad unless it can be accommodated as hand luggage. If you cannot take a telescope on the plane with you, then leave it at home!

Observing Conditions and Weather

Observational astronomers have two main enemies, neither of which we have any control of, yet they continue to fray our tempers and spoil our views. They are, in no particular order, light pollution and cloud, or bad weather in general. Light pollution is becoming worse, a fact that cannot be denied, and on a global scale there is not a great deal we can do about it. However on a local and even national level, campaigners have set up societies to bring awareness about this problem and hopefully halt or even reduce its effect on our starry skies. The Campaign for Dark Skies, is one such group that lobbies governments to try and educate councils, builders and developers to use effective lighting that illuminates the buildings and car parks, and not the sky. These forms of astronomy friendly lighting actually reduce energy usage, and reduce light pollution, so in effect everybody wins. But as with any campaign like this, it is a slow process, but progress is being made. Light pollution comes in several forms, from domestic and local street lighting, to supermarket, motorway and industrial lighting, and is especially prominent in populated cities, where an orange or yellow skyglow can be seen radiating out into the sky from many miles away. Most of this extraneous light energy is completely wasted and should and can be contained, but only time will tell if future generations will truly know what a dark, star-studded sky looks like.

Our second nemesis is poor weather, in particular cloud, and there is absolutely nothing that you can do about this problem other than plan around it, by keeping an eye on local weather reports and occupying yourself with non-observing tasks. Several websites are available that can give fairly accurate local weather reports, even based on Postcodes or Zip codes, and they tend to be more useful than the typical coverage found in newspapers and TV. Use a web browser to search for a relevant site within your own location. It is still possible to observe with scattered cloud, and certainly in the UK this is quite a common scenario, but very often the object you want to locate is behind cloud, yet the clear parts of the sky have nothing of interest! A similar condition always occurs when you buy a new telescope – clouds always congregate for several weeks whenever I purchase new equipment! However, some of my clearest night skies have occurred directly after a period of heavy rain, which appears to clear the atmosphere of dust and pollutants, so bad weather is not always negative.

There are two main atmospheric factors that affect our ability to resolve astronomical objects, "seeing" and "transparency". The first term, seeing, is basically a description of the how "steady" the air is, based on the telescopic views obtained. Air currents at different temperatures and altitudes within the atmosphere cause fluctuations in the transmission of the light, which cause the image to resonate and flicker at the eyepiece. During hot weather, these effects can be exacerbated by hot currents of unsteady air rising from the ground, or even trapped inside the telescope tube itself resulting in further unsteadiness of the object being observed. Seeing conditions can affect observing quite considerably, so the level should be recorded in any logbook or observation form using, for example, the popular "Antoniadi scale", which is based on the following classification: –

I – Best seeing
II – Good Seeing
III – Typical seeing
IV – Poor seeing
V – Extremely bad seeing

Transparency is also an atmospheric condition that describes the clarity of a night sky, and is affected by extinction, or the dimming of an object in relation to the observer's horizon. Transparency is usually measured by determining the faintest star that can be seen with the naked eye, generally at a high altitude away from the horizon, such as the celestial Pole or the Zenith (directly overhead). Some observers also record this value as the "magnitude limit" of their location, which is the faintest star that can be resolved without optical aid on a given night.

Most astronomers will have noticed that stars appear to "twinkle" in the night sky, especially those that are closer to the horizon. This shimmering effect, called scintillation, is also due to turbulent layers in the Earth's atmosphere that cause rapid variations in a stars luminosity due to refracting light. Any objects viewed in these locations will suffer badly in terms of visibility and should be avoided.

An often-overlooked "light polluter" of deep-sky observing in particular is actually a popular target for some observers, the Moon! At full phase, the moon shines at magnitude −12.7, the brightest object in the night sky, and can swamp out all but the brightest star clusters. So, whenever possible, arrange your observations around the time of New Moon when it is completely unlit by the Sun from Earth, or at Crescent phase when its light emission is not so drastic.

Analogue and Paper Resources

Planispheres may be simple and fairly crude, but even in today's modern computer age they provide a quick and easy way to determine which constellations are visible at any particular time. Although they only display a few brighter stars and deep-sky objects, a quick spin of the dial gives an instant overview of the night sky without the need for electrical power or complex set-ups. The Philip's Planisphere, a good purchase, is available in a number of sizes, and a variety of longitudes from good bookstores, but make sure you get the right one for your location.

Paper-based Star Atlases and finder charts are still very popular, even with the advent of sophisticated computer-based programs, probably because they are "ready to go" in an instant, cost far less than a complete PC system, and the clarity and detail of their charts are difficult to surpass. For fairly basic observing, "Norton's Star Atlas" 20th Edition, edited by Ian Ridpath is a useful tool, and also contains a wealth of general astronomy and observing data that all amateurs will find useful. More advanced and detailed atlases include "Sky Atlas 2000", by Wil Tirion and Roger Sinnott, which comes in a variety of styles, for desk and field use and has an extremely detailed yet clutter free format. The "Uranometria 2000" also from Tirion, and the "Millenium Star Atlas", again from Sinnott, are also worthy contenders, but their multi-volume contents mean they are more expensive. Both editions contain a vast array of objects suited to the intermediate observer or particularly with the Millenium Atlas, the more advanced astronomer.

For use in the field, the laminated, wire-bound or single-sheet versions are recommended to avoid damage from dirt or moisture, or alternatively you can photocopy relevant charts (for personal use only!) and scribble down comments and notes on these without damaging the originals.

Although there are many books available on general deep-sky observing, there are few that deal specifically with star clusters. One notable exception is "Star Clusters", by Brent Archinal & Steven Hynes, which is a highly regarded catalogue and reference book. General deep-sky books that you may find useful include "The Messier Objects" by Stephen O' Meara which deals specifically with the Messier catalogue, and "The Caldwell Objects" which covers the Caldwell observing list and is by the same author. "The Deep-Sky Observers Year" by Grant Privett and Paul Parsons is another great practical observing guide, as is "Visual Astronomy in the Suburbs" which gives effective advice on observing under less than ideal conditions, and was written by Antony Cooke. Author Philip S. Harrington has also written an excellent book on general observing, "The Deep Sky: An Introduction" which is especially useful for beginners.

Several monthly magazines are available in the UK and USA that deliver practical advice on observing and help with purchasing decisions, along with topical discussions on a great variety of subject matters. "Astronomy Now" and "The Sky at Night" are British magazines that are very popular, and in America, "Astronomy", and the excellent "Sky & Telescope" magazines are published, all available from newsagents or by subscription, both locally and abroad.

Computerized Resources

The computerized planetarium market is very competitive, and there are many good programs available to cover the needs of most amateur astronomers, regardless of their level of expertise. It would be almost impossible to review or even mention every product in the marketplace, so instead I will briefly discuss what I believe are the most useful products, and especially the software packages I have first hand knowledge of. As most of these programs have similar feature sets, I will try to highlight any particular strengths where applicable. First up is "The Sky," from Software Bisque that is a very sophisticated program available for Windows and Macintosh platforms, and is available as "Student" or "Full Versions." The Sky has an easy to use interface, with a pleasing screen display, but it is particularly good for printed charts, and for this reason it was chosen to create the finder charts for this book. Another excellent program is "SkyMap Pro," from Chris Marriot, available for Windows only and in full or "Lite" versions, it also has an intuitive interface and high-quality charts, and is particularly noted for its extensive object databases and accuracy. Other products that are worthy of further investigation are "DeepSky 2000" from Deep Sky Software and "MegaStar" from Willman-Bell that both have loyal fan bases and have received favourable reviews. "DeepSky 2000" can be supplied as a 2 disc DVD version that has masses of data and images, and "MegaStar" also has a large database library, and is used by many serious amateurs. The benefits of computerized planetariums or deep-sky software are obvious, but include, unlimited and completely customisable star charts, the ability to quickly find objects, and zooming and panning of fields that simply cannot be done with paper charts. The more sophisticated programs also offer telescope control, observation planning and observing list generation.

For field trips or even backyard sessions, all of these packages will run in one form or another on a laptop computer, but for those on a budget or looking for a more compact solution, consider a Pocket PC or Palm PDA package. These miniature hand-held computers can run a variety of astronomy software, that are much more than just "toys." One such product is 2Sky from *www.2sky.org*, for the Palm PDA, which is a surprisingly powerful little program that can display the entire NGC and IC catalogues, has an extensive search facility, and even a Red Screen mode for field use. The Pocket PC mini-Windows system has not been left out, and several packages are available, including "Pocket Stars" which also features the Messier and Caldwell catalogues, search functionality and accurate object positional data.

If you are interested in fully automating your observation targets lists, this can be done electronically within some planetarium packages or using stand alone observation planning software. "Astroplanner" is a software program from Ilanga Inc. that uses an extensive database that can be used to plan, organize and generate observing lists, and keep tabs of your precious observations with an electronic logbook. It is also possible to create personalized target lists directly on the Internet; for example, the NGC/IC Project has such a product at *www.ngcic.org/oblstgen.htm* that can automatically create observing plans by type, constellation and other criteria. Similarly, another online observing list generator can be found at *www.virtualcolony.com/sac*, which is based on the extensive 10,000 object plus SAC database from the Saguaro Astronomy Club.

General Resources

Joining a local astronomy club or a national society is one of the best ways to meet people who share your passion for astronomy, and a great method of sharing ideas and information. From my own experiences, I would advise that you do both and I am sure you will find it beneficial. Local astronomy societies which allow face-to-face meetings and often hands-on practical advice are situated all over the country - a quick search on the internet should find one near you - as most societies are conveniently named after their town or village.

National astronomy societies in the UK include the Society for Popular Astronomy (SPA) which has a magazine, a newsletter and numerous observing sections you can join for a modest fee, and is especially suited to newcomers to the hobby. The British Astronomical Association (BAA) has a similar structure and organization but is aimed at more seasoned or experienced astronomers, and they also produce a good selection of printed material. A specialist organization such as the Webb Society is also recommended, and this club specializes in deep-sky observing, and has a magazine and observing sections. The Webb Society has also publishes a very useful range of books for members and non-members that are well respected.

On closing this chapter, it is worth mentioning the many observing programs and catalogues that can take some of the guesswork out of creating and planning your own customized observing lists. There are a number of readymade general deep-sky catalogues that contain a good selection of open and globular clusters that can be observed in order, or as they take your fancy, so to speak. Famous examples include the Messier Catalogue, The Caldwell Catalogue, The Herschel 400, and the SAC 100, all of which contain a smaller selection of objects than the much more daunting NGC and IC lists. Also, in Chapter 15, a comprehensive observing list of carefully selected objects is presented which should provide a range of targets from very bright, easy clusters through to those which are much more difficult and faint.

Chapter 13

Observing Guide and Techniques

When observing star clusters, we are dealing with two distinct types of object that require slightly different treatment to attain the best results. The loosely scattered open clusters and associations are often brighter and easier to resolve than the tightly packed spherical globulars, which are generally more diffused and cannot always be resolved into their stellar cores. With this in mind, in this chapter I will discuss how to observe these clusters, and introduce some hints and tips that should hopefully help.

Observing Basics/Overview

Before setting of on a cluster observing quest, you need to establish a few basic concepts to ensure you are familiar with the night sky. Take the time to learn as many constellations as possible; this can be an ongoing process from observation, which is the best way to learn to navigate the night sky, or from perusing a planisphere or star charts and committing them to memory. Also, familiarize yourself with some of the brightest stars within these patterns, as these will often be refereed to as celestial markers when hunting actual objects. Learning to orientate yourself and your telescope is also an important lesson which can be practised during the daytime, but traversing and understanding sky directions really needs to be accomplished under the stars.

A good starting point is to work out your position on Earth, and an easy way to do this is with a compass. Once you have established "magnetic" North, the North Star or "Polaris" should be visible in this direction. This star shows the orientation of true North, and is a useful star to use as a baseline, and for polar alignment of the telescope which is discussed later. Polaris is a circumpolar star, so it is always visible above the horizon, and is close to the celestial pole, the point at which the sky appears to rotate and, therefore, Polaris is always in the same part of the sky from the same location. The North Star can also be used to distinguish the latitude of your observing location, which is equivalent to the altitude of this star. Determining the East to West bearing is also crucial as this is the route the stars will follow across the sky during a nights observing, caused by the Earth's rotation.

If you can imagine the celestial sphere as a 180° dome for each hemisphere, with 90° at the head or Zenith, and 45° at the halfway point between the Zenith and the celestial equator, this artificial model can put some perspective on the position of astronomical objects. For example, in the UK, we can technically observe objects down to about −38° and all the way up to +90°, but in the Southern Hemisphere the opposite would apply and in this region, for example the star Polaris would not be visible. A simple way to define angular or celestial distance is to use your hand at arms' length positioned over the night sky, where a clenched fist represents about 10°, and a hand span from little finger to thumb is approximately 20°.

Stellar coordinates for the position and location of objects are measured using Right Ascension (RA) and Declination (Dec), which are equivalent to longitude and latitude on Earth. Right Ascension is in essence, the right to left position and is formulated using hours, minutes and seconds, which represents the Earth's daily rotation from 0 to 24 hours. Declination is basically an up or down measurement and is expressed in degrees, with 0° at the celestial equator to +90° and −90° at the North and South celestial poles, respectively.

To locate a cluster, you will more than likely use several field stars that can be used as pointers or guides to the actual object, and these stars are labelled using a variety of naming conventions. Many of the brightest stars in each constellation have letters from the Greek alphabet assigned to them combined with the constellation name; for example, the star Betelgeuse in Orion is also known as alpha Orionis. Other classifications that you will come across include the Flamsteed numbers, such as 58 Orionis for the same star. The Greek system is also referred to as the Bayer classification.

Greek Alphabet (lower case)

α alpha	β beta	γ gamma
δ delta	ε epsilon	ζ zeta
η eta	θ theta	ι iota
κ kappa	λ lambda	μ mu
ν nu	ξ xi	ο omicron
π pi	ρ rho	σ sigma
τ tau	υ upsilon	φ phi
χ chi	ψ psi	ω omega

Luminosity or the brightness of star clusters is described in terms of magnitude and the basics of this were covered in chapter 1. However, the quoted magnitudes for clusters are based on the total cluster, not any individual star. In other words, the cluster is rated as if it were a point light source and not a diffused object. Cluster magnitude figures can therefore be misleading because a globular cluster could be listed as Magnitude 8, but the brightest star within the cluster might only be magnitude 15.

The apparent size of a cluster, also often stated in catalogues and sky atlases, is based on the angular measurement of an object in relation to the celestial sky, and is measured in degrees, arc minutes, and arc seconds. One arc minute is equal to 1/60th of a degree, and an arc second is 1/60th of an arc minute. For example, the Moon is half a degree or 30 arc minutes in diameter.

Locating Targets

Prior to physically trying to locate a star cluster, you need to "mentally" locate it first using a planisphere to ensure the constellation the object resides in is visible. Once you have established the object is above the horizon, look up the coordinates of the object, and then locate it on a star chart. Alternatively, if you have a software program, this should indicate which constellations are available, and a search function should enable you to locate the object you wish to observe. The planning stages of locating objects "on paper" or digitally were discussed in Chapter 12, which you may wish to refer back to.

Once you have designated a virtual target for observation, it is time to go out and locate it in the real sky. Some objects are particularly easy to locate, as they are positioned close to a bright asterism or star pattern; for example, M13 the Hercules globular cluster, which is almost dead center between two bright stars of the "keystone" asterism. However, most clusters have a less obvious placement, so you need to take some form of star chart or finder chart into the field with you. It does not matter whether this is a sky atlas, a printout from a software program, or even a page from an astronomy magazine, as long as it shows the object clearly and has some bright stars nearby we can use as pointers. Before actually looking through the telescope, and to obtain the general observing area, I find it beneficial to sweep the target with a pair of binoculars so you get an instant wide field view of your bearings. Later, you can also refer back to binoculars for a wide angle perspective if your target goes astray. You can now point your telescope in the approximate direction with confidence, lining the optical tube up with a nearby bright star plotted on your chart. One of the best and easiest ways to locate the object in question is by "star-hopping." Using the chart as a reference and viewing through your finderscope simply follow, one by one, the line of stars that lead from the bright reference star to your destination, rather like Hansel and Gretal following the breadcrumbs! If you have performed this sequence correctly, you should be close to the target object when you look through the main telescope.

There is no right or wrong way to define this star-hopping order; it can be a simple direct path, or a convoluted trail, as long as it gets you to your destination, but it is wise to select as many bright easily distinguishable stars as possible. A unit finder certainly helps to get a general fix on orientation, but an optical finder is often better for star hopping as you will be able to discern fainter stars. Star hopping can be used to track down a great number of clusters, but in barren star fields, pointer stars are harder to find, but there are nearly always reference stars somewhere in the region that can point you in the right direction. You should also bear in mind that depending on what type of telescope and finder you use, the optical image will inevitably be inverted or flipped in relation to your star chart, so ensure you take this into account when moving around the sky.

To aid in keeping your target centered in the field of view, RA & Dec Drives are not essential but extremely useful. At high power, celestial objects stray from the center of view in a matter of seconds, but using drives counteracts the rotation of the Earth, so the object appears to remain stationary. Once the object is positioned, the drives will keep it centered for up to a couple of hours, depending on accuracy of the drives themselves, and the initial setup of the polar alignment. Alternatively, hand-operated RA and Dec knobs are supplied with many telescopes, and these can be turned to keep the object in the field of view, though some considerable "twiddling" may be necessary at high magnifications. Polar alignment is an important procedure if you require accurate tracking, or if attempting astrophotography or CCD imaging, which are discussed in the next chapter. When performing alignment, the tripod must be levelled first, which can be achieved with bubble or spirit levels incorporated into some mounts. The latitude of your location can then be set on the RA axis, and the Dec axis should be set to 90°. Both axis' can then be locked into position, and if necessary, the Latitude can be adjusted until the Pole Star is centered in a low magnification eyepiece. Once this has been set, the mount will accurately track the sky using the Right Ascension drive or hand control. Several equatorial mounts are supplied with a built-in "polar alignment scope," which can assist in aligning Polaris, and these can also be retrofitted on some models.

All RA and Dec. dials have guides marked on them called "setting circles," which are seldom used by many amateurs. This is mainly because, on most equatorial mounts they are quite small so their increments are crude and the measurements are therefore approximate, but they can be used in various ways to good effect.

There are several ways of locating objects with setting circles, but using a guide or reference star is probably the easiest and most effective. Simply point the telescope at a bright star of known RA and DEC near to your target, and adjust the RA dial to match the coordinate of this star. The Dec dial cannot be adjusted but should indicate the correct value if the mount is correctly polar aligned. From this reference point, you can use the circles to find objects that are in the same part of the sky, simply by moving the telescope until the readings match your targets coordinates. However, if you move to a different part of the sky, inaccuracies can occur, so you must recalibrate from scratch with a new reference star. To obtain an even more precise calibration, digital setting circles can be obtained for equatorial and Dobsonian mounts that use encoders on both telescope axes' to provide continual, precise digital readouts of your position in the sky. These devices also include an integrated deep-sky database that can be used to zero in on your target.

Modern innovation has brought the computer revolution to the telescope and "robotic" or Goto telescopes are now common. There are two trains of thought for this type of technology. The scope will find objects automatically which saves time and enables the user to see more objects during an observing session. With ever-degrading skies and limited social time for many, this seems like an ideal proposition. The downside is that you never really learn your way around the sky, or how to find objects by star hopping, which is often half the fun. Goto technology is still quite expensive, and if you can do without it, you may be better off spending the money on superior optics. One exception to this thought is for observers who perform CCD imaging, as top of the range Goto scopes have extremely precise tracking – though in this scenario, the electronics are being used to "stay on" rather than "go to" an object.

Observing Techniques

Observing should be enjoyable, whether you are observing alone or sharing the experience, but failure to resolve detail in a cluster, or even find the object in the first place can lead to frustration. Several of the techniques mentioned here should remedy some of these potential problems.

Dark adaptation can make a serious improvement to an observing session and should be taken seriously. Extinguish as many external and house lights as possible, disable any security lighting, and depending on your relationship with neighbors, you can kindly ask them to do the same. Whilst setting up your telescope, it is wise if possible to reduce the amount of light as soon as possible, even assembling equipment in the dark with red light, so dark adaptation can begin. In

Chapter 10, I commented on the eyes, ability to adapt to dark conditions using the night-sensitive rods, but these take time to become fully operational, so the sooner you begin dark adapting the better.

Using a shroud or hood over the eyepiece assembly can also help to reduce stray light, if this is severe in your own location. To speed up the dark adaptation process, try staring at the dark sky for several minutes, or close your eyes to get those rod cells working overtime. Some observers have even been known to wear sunglasses indoors before venturing out into the night sky to aid their night vision, others might suggest this is taking the subject too seriously! However, once your eyes are fully adapted, you should be able to see up to 2 or 3 magnitudes fainter, which makes a massive difference. Whilst observing, use a red light or torch which will not affect your newly acquired night vision and try not to re-enter any lit areas, as your eyes will quickly revert back into daytime mode, and they will require a further 30 minutes to fully resume it.

It is important to take your time whilst observing. It is not a race, and there are no medals for the person who can see the most objects in one sitting. Instead, it is better to savor each cluster, and spend some time getting to know it. Even if you completely fail to find an object, you can simply move on to another target and try again for the unsuccessful cluster on another night. Repeated observations of the same objects can also improve your original experiences and you are likely to render more detail each time you revisit an object. Whilst observing you may also want to try looking away occasionally, as fresh eyes can often focus better and moments of good seeing can improve the view.

When you believe your telescope is in the correct place but you are struggling to see anything, or you cannot discern any detail in an observed cluster, you can try the following methods. Applying slight movement or tapping the telescope tube can cause some "invisible" objects to suddenly pop into view, or previously unseen detail can suddenly become apparent. This is because our eyes seem to respond better to moving objects rather than stationary ones in low-light environments. Similarly, to confirm the presence or absence of an object, move the telescope away from the cluster and let it drift back into the field of view, and the eye should detect it.

Looking at a telescopic target directly is comfortable and natural, but the central section of the eye is the least effective part in low light conditions. Averted vision can aid in detecting faint detail, and involves looking off to one side, using your peripheral vision, in the outer regions of the eye, which is more sensitive in the dark. This technique sounds odd, and requires practice to perfect, but it really does appear to work. Perhaps this ability stems from the days when our ancestors had to be aware of predators lurking in the dark, just out of view? You may also notice that certain parts of your eye can resolve more detail than other areas, or that one eye can discern more than the other – test this the next time you are out observing.

Binoculars are a useful tool for scouting clusters as they are light, easy to use and completely intuitive when locating an object. Another benefit is that they use mirrors and prisms to counteract any optical inversion or mirroring, so a binocular view represents objects exactly as they appear in real life. Telescopes do not possess this quality and all types deliver a reorientated image in comparison to the original image. Reflectors invert the image, so it appears "upside down" but this is not of major concern, as you can simply rotate your star chart to match the view at the eyepiece. Refractors, on the other hand, flip the image left to right, providing a mirror image, which is more confusing, but many computerized planetariums can flip their chart output to compensate for this effect.

The cardinal points North, South, East and West can appear in a multitude of directions depending on how the telescope is positioned, with optical inversion and mirror images confusing this even further. To orientate the telescopic view, allow the image to drift across the field of view; the direction of the stars movement indicates West.

This "drift method" can also be used to calculate the telescope's field of view with any given eyepiece. To calculate the field diameter, determine the length of time it takes for an object to traverse the telescopic view and multiply this by 15. For example, if the cluster takes 5 minutes, multiplied by 15 gives 75 arc minutes, or 1.25°.

If you experience any problems acquiring perfect focus on a particular cluster, one solution is to nudge the telescope over slightly to a brighter star near the field, taking note of this direction, then focus on the alternative star instead. You can then orientate the telescope back to the original cluster and your focus of the object should be much improved. Two of the best tips are, in fact, patience and perseverance! Locating and observing some of the faint or obscure clusters can be difficult, so do not be too hard on yourself if you fail the first time. Several years ago, I once took over an hour to locate a single object, then found I had located the wrong one! Keep trying, even if you initially struggle, and eventually you will succeed.

Observing Open Clusters and Associations

Open clusters can be observed all year round, but are especially prominent in winter skies, with some of the brightest examples on show, such as the Pleiades and Hyades in Taurus. During winter, there are many other excellent open clusters visible in several constellations including Gemini, Auriga, Monoceros, Cassiopeia and Puppis. In the clear skies of summer, open clusters are also plentiful in Sagittarius, Scutum, Cygnus and Vulpecula, to name a few. Open clusters can also be observed during the spring and autumn but the number and variety of objects is more limited.

Open clusters tend to be spread over a large area and require a wide field and low magnification to be completely contained within the telescopic view. Using a high-power eyepiece will simply over enlarge the cluster and you will not see it in its true glory.

A long focal length telescope (higher than f6) can produce a similar result, so a fast, short focal ratio is preferable for these clusters. Binoculars are also an excellent choice for observing open clusters, especially the more luminous groups. In fact, several look better in binoculars than telescopes of any type or size. Once you are in the approximate region of your target object, using the methods described above to get to this position, you should be able to see the cluster at low power. If your position is correct and you cannot detect the cluster, pan around the area and look for clumpings of stars that appear more condensed or brighter than other field stars. Faint clusters are more noticeable when the telescope is moving and the eye can often suddenly detect an object in this way.

Certain clusters can still be elusive and blend into the background stars, so if necessary, confirm your position.

Crowded fields where there are many stars of similar magnitude can also be problematic in determining an object but multi-clusters are not common, although they do appear, such as the double cluster in Perseus and close clusters in Auriga, which contains over 40 objects. When they do occur, however, they are not tightly crammed like galaxy clusters, so you should not have any difficulty in unravelling one cluster from another. Whilst observing open clusters, make a mental note of the clusters size in relation to the eyepiece field of view and count the number of stars visible using different eyepiece configurations. If the cluster is dense or highly populated, you can simply estimate the number of stars based on a particular region. Published data on star numbers within open clusters vary dramatically, as quoted figures can be based on the actual number of stars from a professional study, or from stellar counts visible in amateur size telescopes, so take this into account when estimating member star numbers.

Bear in mind also that some of the apparent cluster members are in reality foreground or background field stars, but even professional astronomers have difficulty extracting true cluster stars. This does not matter and you should observe and record whatever you think is relevant, as no two visual observations are identical anyway and most observers see objects slightly differently.

Various colors can also be observed in open clusters and several bright examples take on a reddish hue which emanates from the red giant stars within the cluster, a good example is our old favorite the double cluster. Blue stars are also prominent in certain clusters; the most obvious example is the Pleiades, which has a notable blue cast from its hot young stars. Color is generally more evident when using direct vision, whereas subtle detail is more apparent through the averted vision technique. Also look out for multiple and double stars that can appear both in tight groups, and in widely separated associations, which are made up from combinations of optical pairs and genuine binary stars.

Any other details that you notice whilst observing are important and should be recorded, such as the differences in color and brightness of the component stars, and any obvious patterns and shapes that stand out. These can be used at a later date when recording and logging your observations, and for comparisons with existing images and catalogues to confirm your observation, which will be covered in detail in the next chapter.

Nebulosity is associated with certain open clusters, such as NGC 7510 in Cepheus, but it is not always obvious whilst observing, whether this is gaseous nebula that is emitted or reflected from the cluster itself, simply refracted light, or star glow from the member stars. Clusters embedded within reflection or emission nebula make interesting targets, and include objects such as NGC 2244 situated in the Rosette Nebula, and M16 which is associated with the infamous Eagle Nebula. If you suspect any hint of nebulosity whilst observing, make a note of this so that you can verify this with existing catalogue data.

Bright open clusters are visually very appealing and impressive, though some of the fainter ones can be a little nondescript but even these are well worth tracking down. As many stellar associations are grouped around or near to known open clusters, they can be observed using similar methods as described. However, as they are much larger in apparent size, extremely wide fields of view are a necessity. To fully encompass these stellar groups, a rich-field telescope or astronomical binoculars will help a great deal.

Observing Globular Clusters

Globular clusters are stunning objects from a visual observing perspective, and are arguably one of the best deep-sky targets for amateur astronomers. Several of these are visible with the naked eye, although mainly in Southern Hemisphere; for example 47 Tucanae and Omega Centauri that are both bright and extended. Some globulars are visible with binoculars; for example, M13 in Hercules which appears as a "fuzzy" star, and is the Northern Hemisphere's finest example. Although there are a number of globular clusters on show throughout the year, the best time to view these spherical giants is during the summer, when the Milky Way does not obscure the view. Ophiuchus is a constellation replete with globular clusters, and they can also be found in Serpens, Hercules and Sagittarius. During spring, there are also a fair few globulars on display in Canes Venatici, Coma, Hydra and Bootes, but in autumn and winter, they are few in number as we are looking towards the wrong part of the galaxy. There are, however, some decent globulars in autumn, within Pegasus and Aquarius and a few lonely clusters in Lepus and Columba during winter. Some of the best northern globulars include the Messier objects M2, M3, M5, M15 and M92.

Globulars vary quite considerably in star content and density, such as the bright, loose globular M107 in Ophiucus, which is a class X object and is easily resolved in small and medium telescopes into its stellar components. Rich, compact globulars like M2 in Aquarius, a class II cluster, require higher magnification and large apertures to even begin to resolve the clusters to their stellar cores. It is quite useful to refer to the Shapley system of classification, discussed in chapter 4, which gives a good indication of the likely possibility of resolving a globular into its constituent members. The ability to resolve such detail is largely dependent on the aperture of the observer's telescope, and although much can be done with a 4" wide field refractor, an 8" reflector, or similar size instrument is the optimum choice. Some of the faintest globular clusters, such as those from the Pal (Palomar) and Terzan catalogues (Ter) require telescopes in the region of 20" and larger to event detect them, which is one of the reasons why massive Dobsonian telescopes have become so popular.

At first glance, many globulars appear superficially the same in terms of size and structure, but on closer inspection you will see that they all have quite distinct physical and visual properties. Some are tightly packed spherical aggregates, whereas others are scattered and appear disrupted to some extent. Typically, the most luminous globulars are 20 to 30 arc minutes across, and appear larger than some of the fainter objects because they are somewhat closer and consequently easier to resolve. The ability to scrutinize stellar detail within a globular cluster is, therefore, rather dependent on the object's visual magnitude, though aperture is a contributing factor. Using the highest powers available, some clusters are difficult or even impossible to resolve visually in typical amateur telescopes, for example M72 in Aquarius, and M53 in Coma, because they are distant, and therefore small and faint, generally between 4 to 8 arc minutes in diameter.

To combat these limitations, the advent of CCD and Digital Imaging has become a powerful weapon in the arsenal of amateur equipment and is covered in detail in the next chapter.

To observe a globular cluster start with a low power, wide field, and look for a blurred or bloated star that should stand out from the field stars within the view.

Globulars are quite diffuse objects, often resembling a galaxy, or even a comet at low power, but they should appear noticeably different from surrounding stars. If you suspect the presence of a cluster, increase magnification to medium power and some mottling or partial structure should be evident, confirming your suspicion. High magnification can be used to really open up the cluster's composition and resolve down to the member stars on brighter globulars. Experimenting with different eyepiece and barlow configurations can tease out further detail, but you will eventually reach the point where too much power actually detracts from the quality of the image. Unsuccessful location of a cluster is common, and if this occurs, try moving the telescope slightly as the globular may be just outside the field of view, otherwise re-examine your charts and try again by star hopping or with setting circles.

A surprising amount of detail can be detected at high powers, and many globulars show evidence of texture and shapes, clumping of stars and mottled structure. These effects are prominent throughout the cluster but the globulars core is notoriously difficult to penetrate, though digital "imagers" have performed wonders in these areas. Chains or strings of stars have been described by many amateurs and these are more pronounced on the clusters outskirts where they appear to fade off into obscurity. In large apertures, dust lanes that obscure the underlying stellar material can be observed, so look out for any areas that appear lacking in stars, especially in the outer regions. One of the most interesting stellar "absences" is a distinct "Y" or propeller-shaped dust lane in the M13, the Hercules Cluster, but this is difficult to see even with large telescopes. Once again, digital imaging comes to the rescue and amateur CCD's have been able to pick out this structure quite clearly.

Although globulars tend to be spherical in nature, some contenders are considerably asymmetrical in structure; especially those close to the Galaxies bulge, where they are more affected by tidal forces. When observing these types, see if you can define this configuration in terms of orientation and extent. The subject and existence of visible color within globular clusters is controversial, but there are a number of reports of varying shades and hues that have been observed by amateurs. With such diffuse, low surface brightness, objects this color cannot come from individual stars, and is either an overall color cast from the combined stellar members or some form of visual illusion. Many observers, myself included, only see shades of grey within globulars, but if you can see color then you should certainly record this fact.

Chapter 14

Imaging and Recording Objects

During the last decade, and particularly in the last few years, digital technology has encroached into all areas of amateur astronomy, with computerized telescopes, "goto" mounts and even GPS satellite-based alignment now commonplace. This growing trend is particularly evident in the field of imaging and recording, which has been completely digitized and has revolutionized the hobby for many amateurs. While this is a welcome progression, allowing astronomers to detect and record objects beyond the realms of previous generations, there is still a place for visually observing and making sketches at the eyepiece. Astro-photography, on the other hand, has largely been superseded by the advent of digital imaging, but can still provide excellent results with only a modest outlay.

Visual Observation and Sketching Clusters

Regardless of the aperture or cost of your telescope, unless you have learnt the "art" of observing, this will have little effect on your results. Learning to really "see" star clusters takes time and experience, and there is no better way to improve these skills than by visually observing, and drawing what you see. Sketches at the eyepiece have little or no scientific value, and nearly always contain a certain amount of artistic license, as most observers visualize and, therefore, render their drawings differently. However, none of this matters one iota, because these drawings are "your" personal recordings of these objects, and as you observe and sketch in the details you become a better observer. Artistic skills are not a prerequisite, as long as you make the drawings as accurate as you can in terms of position and approximate size. Readymade observing forms are available from most local and national astronomy societies for sketching purposes, you can also design your own if you prefer, or simply sketch on plain paper, and transfer these to proper reports at a later date. Other than paper, the only other equipment you need is a soft pencil, a rubber, and a clipboard to hold everything in place and some form of red light, such as a torch or clipboard light. When sketching astronomical objects bear in mind that you are essentially creating a negative image, in other words, the brightest objects will be the darkest in your drawing. However, a pleasing result can be obtained by scanning these drawings after observation and inverting them using an image-processing program, so that they represent the real sky, with white stars on a black background.

Once you have located your target, begin the drawing by pencilling in any bright field stars to orientate and scale the sketch. For open clusters, you should render the bright stars first, adding fainter members later on. To indicate different stellar magnitudes, bright stars should be drawn larger, as circular "blobs," whereas the faintest stars will simply be a dot of the pencil. I find it useful to start from the inside of the cluster working out, but you can build the drawing any way you like. If the cluster is very rich you do not have to add every single star, just an approximation will suffice, and for nebulosity you can simply indicate this with a line which can be shaded and blended later. Any double or multiple stars that you suspect should always be included, and will help you to correlate your drawing with those found in star charts or observing guides. If you make a mistake, simply rub it out, but keep swapping between your sketch and the eyepiece to ensure you do not stray too far from the appearance of the object being drawn. Some observers use a ballpoint pen to indicate the stars as it is easier and quicker to create a dark mark, but I prefer pencil so I can quickly eradicate all my errors!

Globular clusters require a similar treatment, so add in any bright field stars before you begin sketching the actual cluster. Because globulars are generally faint and diffuse, you may not be able to resolve individual stars, but you can indicate the general size and shape of the cluster by applying light amounts of lead, and then blending and smudging the pencil to create a spherical glow. This will be darker towards the core, gradually blending into much fainter shades at the clusters extremity. However, if you can resolve stellar components, indicate these to the best of your ability, and try to accurately position any chains, strings of stars, or other fine details that may be evident. To orientate your drawing, take notice of the direction in which the telescopic object is drifting, and mark this on your sketch as "West."

Several different eyepieces and magnifications can be used for a single sketch to create the final composite image, so experiment with different combinations of low power to define overall shape and structure, and high power to tease out the details. If your sketch is produced over a period of 20 minutes or more, throughout this process the seeing conditions will fluctuate considerably, so be mindful of those rare moments of excellent seeing when additional detail can suddenly become apparent.

When you have successfully completed your sketch, go back to the eyepiece and compare the finished drawing with the cluster at the eyepiece, and study this for a few minutes. Wish fresh eyes you may see some extra detail, or wish to make some subtle alterations.

Astrophotography

Unless you have an accurately polar-aligned telescope and a reliable RA drive of some sort, astrophotography or any digital imaging for that matter is best given a wide berth. Using conventional analogue films, the exposure times required to image clusters, especially the globular variety are quite lengthy and only a critical setup would provide any suitable results. If you cannot precisely track these objects you will either image severe star trails and image streaking, or worse still, a blank image.

Astrophotography was, until recently, the standard way to record celestial objects, but digital methods have rendered the process almost obsolete. In fact, several film manufacturers have ceased production of some monochrome films used in astronomy that were once the de facto standard. There are, however, still die hard fans of film-based imaging, and one great advantage is the relatively low setup cost in comparison to digital technology. Basically, all you require to perform astrophotography, once your telescope and mount are in order, is a manual 35mm SLR camera, a means of attaching it to the scope, which can be obtained from most astronomy shops, a cable shutter release and some film. The camera itself must be fully manual or have a manual override, so that long exposures can be achieved, but in this modern age, manual cameras can be hard to come by. There are a few manufacturers that still sell these cameras, as advertised in astronomy magazines, but they can be picked up in second-hand stores, or car boot sales for a token fee.

A popular method for setting up a photography system is "prime focus" where the camera sits in place of the eyepiece, attached with an appropriate adapter, but this can cause problems when trying to focus, as the eyepiece and camera will not reach focus at the same point. One solution to this dilemma is an off-axis guider, or flip mirror which directs most of the light path to the camera, and a small proportion to a separate eyepiece to achieve proper focus. This item can also be used to keep an object in the field of view, which improves tracking. Alternatively, you can piggyback a wide-angle telephoto lens onto the main telescope and attach the camera to this lens, which is simply focused at infinity and provides a great way to photograph wide-field star clusters and associations.

It is still possible to obtain specialist mono and color films that have fast emulsion speeds up to 3200 ISO but many amateurs stick with readily available "household" films such as color negative 400 ISO, which can provide good results, and can be processed at any high-street lab. Specialist films require more complex development and some amateurs process their own films but this increases the cost dramatically. Exposure times for astrophotography vary dramatically, depending on film speed and aperture but can be anywhere from 10 minutes for a bright object, to several hours, where some advanced amateurs shoot separate exposures through colored filters which are then combined to create a color composite image.

Some issues particular to astrophotography should also be considered; for example, the dynamic range of film is limited so images that are correctly exposed in mid tones can be completely washed out in the highlight, or brighter regions. Another downside is that you cannot obtain the instant results achievable with digital technology, and until the film is processed and developed, you cannot be sure you have obtained a desirable exposure. Considering the low cost of a basic system, this is an ideal entry into astronomical imaging, and most of the skills that you develop, and problems you encounter will set you in good stead if you decide to move on to digital imaging. However, if you prefer instant results and can justify the cost of a CCD camera, in the long run, this is probably a better alternative.

CCD Imaging

When CCD (Charged Coupled Device) technology first appeared on the amateur scene, it was very expensive, bulky and limited in functionality. This is certainly not the case today, and although they are still costly, prices have dropped but the quality and ease of use is much improved. CCD units are normally used at prime focus, replacing the eyepiece, and off axis guilders can also be used effectively with these devices. Several CCDs have built-in auto guiders that can automatically track a bright star whilst imaging an object, and send instructions to a computerized telescope to make adjustments, keeping the object dead center in the frame for hours at a time.

Many different models are available by several manufacturers, from entry-level mid resolution monochromatic models, to "cooled" full color high-resolution devices. Cooled CCDs are kept at a constant temperature, which improves the quality and reduces the amount of "noise" and other artefacts that can affect image condition. Resolution of a CCD device is based on the amount of detail that the unit can record per pixel, and is usually 12 bit or 16 bit, with higher "bit per pixel" ratios delivering improved results.

The real beauty of CCD, and all digital-imaging processes is the fact that you get instant feedback and results, and no wastage – if an image is poor quality you simply erase it from the computer's hard disk. Unfortunately, many CCD devices have a relatively small field of view. In entry level models, this may only be 4 or 5mm square which is tiny in comparison to conventional 35mm film, but compact objects like globular clusters will fit into this field pretty well. For large wide-fields, such as extended open clusters or associations, larger wide-field CCD models can be obtained but these are still expensive. A cheaper alternative, especially for SCT telescope users, is to purchase a focal reducer, which can convert an f10 system into an f6.3 or even a f3.2, resulting in a much wider field of view, depending on which model you use.

In comparison to film, CCD units have an extended dynamic field, because digital imaging is linear and more responsive to light, and less likely to overexpose saturated areas. Monochrome models are more sensitive than color devices, and this is one reason why many deep-sky observers purchase mono cameras, and planetary observers purchase color models. Single-shot color CCDs record a single color image and the more expensive Tri-color cameras take three separate red, green and blue exposures that are combined electronically. Generally, color can be seen in many deep-sky objects, but as discussed in the previous chapter star clusters do not have overwhelming color like nebulas. One shot color cameras will require a long exposure or many short "stacked" exposures to obtain any decent color deep-sky images, and unless the tracking is spot-on, even combining these images can be problematic, leading to washed out stars. From typical mediocre skies, long exposures are not ideal due to intermittent cloud, variable seeing and transparency conditions, so a mono CCD is more than adequate for our purposes, is more sensitive and can image the deep-sky in a much shorter time.

In terms of CCD exposure, this is a fraction of the time required compared to using conventional methods, and many of the amateur CCD images in this book had exposure times of only 30 seconds and, at the most, a couple of minutes.

Even from a light-polluted suburban site, a CCD can render stars down to 20th magnitude, in a 10" telescope, which is comparable with professional plate surveys using massive telescopes, from only 10 years ago. CCD systems are not without problems and long exposures can introduce "noise" which is comparable to film grain. Flare and glare from external light sources can also impair image quality, but most of these issues can be resolved with image processing. The high cost of a CCD system is also a major issue, and when you add on the expense of a PC it can be prohibitive, especially when these items can cost more than the telescope, mount and accessories put together. However, if you are serious about imaging star clusters, and can warrant this kind of expenditure, CCD imaging will certainly deliver the goods, as countless amateur images have proven.

Digital Still Cameras

Digital cameras, specifically SLR models, are becoming very popular in astronomy but they are still quite expensive, although they can double up as a fine instrument for general photography. Only models with manual controls are suitable for deep-sky work, general low cost "point and shoot" digital cameras are not appropriate as they have few or no manual override functions and their shutter speeds are too fast. The chip on these entry-level cameras are not sensitive enough to cope with the low light conditions typical of star clusters, and several low end digital cameras also incorporate CMOS chips, not true CCD chips which are less sensitive and are found in many webcams.

Digital SLR cameras in particular are defined by their removable lenses, which means the telescopic image is delivered straight to the imaging chip, as with CCD imaging, and is not subjected to a convoluted light path. Prime focus imaging is equally useful for digital and conventional photography, but fixed lens cameras subject the image to unnecessary optical "processing" and are difficult to mate with a telescope, so with these cameras the only suitable method is to marry the camera lens to an eyepiece on the telescope. Focusing and image vignetting are only two of the problems of using this "afocal" procedure, and therefore, the method and results are undesirable.

Because digital cameras are designed for consumer use, and not astronomy, they are versatile and easy to use and have a much higher resolution than specifically designed devices, with 5 million pixel (mega pixel) cameras now common. A major plus point is their "all-in-one" product status, so they do not require the extra expense of a dedicated PC or laptop computer, and setup times are easier and quicker than with a CCD unit. Observations are simply recorded on to the camera's memory card, and with a suitable card reader, these can be copied onto a computer at a later date for inspection and image processing. Electronic "noise" generated by heat from the camera can be a problem with these devices, as they were not designed for taking long exposures in low light conditions, whereas CCD units are specifically built for this purpose. As a digital camera warms up, the noise increases, but as observing sessions are often done in the cold dead of night, the severity of the condition is reduced. Turning off the camera when not in use, and disabling the LCD viewing screen further reduces thermal buildup to counteract noise. High-quality digital SLRs have noise reduction modes to reduce these artefacts, and many CCD units solve these noise problems to some extent with integrated cooling devices, such as fans, heat sinks or even cooled liquid.

Digital cameras generally only allow maximum exposure time of a minute or so, which should enable many clusters to be imaged, but how can you increase the cameras sensitivity further? By taking multiple short exposures of the same object, these be electronically combined or "stacked", resulting in a better image than one created from a single long exposure. As with CCD imaging, focusing a digital camera can be tricky as the electronic viewfinder is useless in dark conditions, and no optical alternative is available on these cameras. In a similar fashion to CCD methods, focusing on a bright field star or the use of an off axis guider can help. Alternatively, you can simply run off a few test frames to check the focus and exposure settings.

Video Astronomy

Nothing compares to the thrill of visual observing as the light from a distant cluster reaches your eye and you are witnessing these objects first hand. A viable alternative that comes a close second is Video astronomy, which provides "real time" imaging of the deep-sky. Developed specifically for astronomy, these units are highly sensitive, low noise cameras that are a far cry from general home video camcorders. Primarily, they are exclusively monochrome devices that use image intensification and frame integration to automatically stack multiple images to form the live video image. Real time imaging allows the observer to view an image before, during and after exposure and the results are surprisingly good, revealing detail visual observers can only dream about. These devices fit into a standard 1¼" eyepiece holder and are controlled from a separate hand box, and although a PC is not necessary, a video monitor or TV is required to display the images.

As a continual display is presented to the user, the experience is more attuned to visual observation than other digital imaging techniques, though some observers may balk at the idea of observing the deep-sky on television. In reality, this situation applies to all digital imaging methods, you are not observing "directly" but you are still observing – it is the nature of the beast. If these technologies allow us to see more, or anything for that matter, in urban and suburban light polluted skies, the trade-off must be that our eye no longer resides at the eyepiece. This has been the situation for many years at professional observatories – you could not use the 10 metre Keck Telescope to observe visually and with all that resolving power, to do so would be extremely wasteful.

Recording images with a video system is a bit painful, the composite video signal can be stored in a variety of ways; for example, onto tape using a video recorder or camcorder, which can be used to show off the images. For image editing purposes, the video must be fed into a computer via an appropriate interface card or frame grabber system. Due to the need for extra equipment required for this type of media, on top of the camera itself, video astronomy is expensive, but if real time observing appeals to your imagination there is much that can be done with these systems. Video astronomy is also ideal for star parties, astronomy clubs and other group activities, as the images can be shared on a large screen or even video projected for meetings and lectures.

Web Cameras

Web cameras cost less than a typical eyepiece and these cheap and cheerful products have been used by planetary observers for several years to produce stunning images of the bright planets. Recently, advances in webcam design and image processing softwares have enabled these devices to be used for imaging deep-sky objects, especially clusters, to great effect. Almost all models are color, but as star clusters are not particularly rich in this area, this is not of great benefit, although some faint color can be perceived on luminous objects. Web cameras are designed to record live or "streaming" data and can obtain up to 60 frames per second, but only have extremely short exposure times of about 1/25th of a second or less. These figures are the key to imaging with these products because you can record multiple images in rapid succession and digitally combine these to create a single frame using image manipulation software. Although off-the-shelf web cameras were originally developed for video conferencing and internet use, they can be used successfully for planetary imaging, but for deep-sky work they do require some modification to enable faint objects to be recorded. If you are electronically minded, and know one end of a soldering iron from the other, these cameras' circuit boards can be adapted to allow less frames with longer exposures to be created. Several websites are available that demonstrate these procedures, but it goes without saying that any manufacturers' warranties would become null and void if you decide to go down this route. Assuming this is beyond some observers' capabilities, several manufacturers have created their own adapted web camera designs which can be purchased commercially, but still cost much less than a conventional CCD unit or high-end digital SLR.

As touched on previously, webcams based on CCD technology are better than their lower quality CMOS counterparts, and there are a few suppliers who distribute such models that appear frequently in astronomy periodicals. Unfortunately you do require a computer to use a web camera setup, but most models are Windows and Macintosh compatible and use standard USB connections to interface with your PC. A dedicated computer or, better still, a portable laptop are ideal, but if you use your home computer it needs to be positioned as close as possible to the telescope, such as near a window or door, or even outside if circumstances permit. Most web cameras are supplied with software, but you will need a specific astronomy package such as "AstroVideo" or "Astro-Snap" to control the camera, which can be downloaded from the Internet. Expert webcam users have resolved stars down to 16th magnitude and many remarkable star clusters, especially bright globulars have been successfully imaged with these cameras.

Digital Image Processing

Astronomy books and magazines are filled with amazing deep-sky images of star clusters, and the Internet is certainly the largest deep-sky gallery in the world with tens of thousands of images. What ties these scenarios together is the growing trend in digital imaging, which can now match and even exceed traditional film-based recording techniques. But the greatest advantage is the ability to enhance and improve this data, turning mediocre observations into works of art! There are many ways in which digital images can be refined, and various software packages that provide these tools to retouch, sharpen, color correct and merge observations, to name a few.

When an original image is taken from any digital device, it is described as "raw" or untouched, and most files are either monochrome greyscales or RGB (Red, Green, Blue) color files that can be supplied to the PC in a variety of formats including JPEG or TIFF. The raw CCD image often suffers from noise or electronically induced artefacts that can be removed using "dark frames." Essentially, a dark frame is an unexposed frame taken with the telescope capped, or the camera shutter closed, at the time of the observation. This image only contains the artefacts themselves, and no other data, and using software, the dark frame can be subtracted from the original image to eradicate noise and other artefacts such as "hot" pixels.

Image stacking is another important software tool and allows multiple images to be automatically aligned and merged to create a final single composite image, that retains the best parts of each individual frame in the completed file. Stacking can combine tens or even hundreds of images to improve the clarity and resolution of many observations; even the color and contrast can be enhanced. Several stand alone software packages are available to perform these tasks, including "Registax" and "AstroStack," again both available from the World Wide Web. Other software programs that incorporate this feature include Maxim DL which is a powerful image-editing program in its own right, and also features camera control, depending on which version you purchase. Most CCD cameras come with some form of basic camera control software, but packages like Maxim offer many more high-end features like image processing controls, image filters and enhanced camera controls for CCD, still, video and web cameras.

General image editing such as sharpening, contrast control or cloning, for removing unwanted pixels can also be achieved using retouching programs such as "Adobe Photoshop" or "Paintshop Pro." Photoshop has a complete suite of editing tools such as "curves," for improving brightness and contrast, "masks" that can be applied to certain areas only and many "filters" that can reduce or remove satellite tracks, lens flare, glare and light pollution. In the right hands and with a bit of practice almost any digital image, regardless of how under or overexposed or blurred it is, can be improved with image processing.

Observation Report and Logs

It could be argued that amateur astronomers should simply go out and just observe objects without recording or logging anything, because it is fun and relaxing. This is a hobby after all, so why bother to put in all the extra time and effort to write notes, fill in forms and make sketches? There are actually a number of very good reasons to get into the habit of recording each and every observation you make. Firstly, making notes and filling out report forms improves your skill levels, by focusing your mind whilst observing, so you tend to pay more attention to objects and determine more detail than you could if just observing casually. By writing down your visual impressions and referencing these later, you also learn more about the cluster in question and about astronomy in general, which can only be a good thing. After a short period of time, you will have created a unique set of records that can be referred to at any time to check or cross reference new, or improved observations of the same object. Observation logs are also interesting to peruse on cloudy nights, and you can compare how your observing skills have enhanced over the months and even years.

I have referred to my own set of observing reports for open and globular clusters frequently whilst writing this book, and am pleased that I took the time to create these records. They really do come in handy and are highly recommended. I find it useful to have two methods of recording observations, the observing report that contains a sketch and all the relevant settings and notes, which I use in the field, and fill in any blanks later. I also find a logbook beneficial, for use after each observing session that simply lists all the objects I have seen with a basic overview, such as "new observation" or "requires confirmation." The logbook can also be referred to more quickly than sifting through pages of observing reports.

Professionally printed and bound observing books and logs can be purchased from specialist bookstores, or you can use the report forms from local or national astronomical societies such as the SPA or BAA. If you join one of these societies and become a member of the appropriate observing section, the section directors will normally supply these forms free of charge. You can also download report forms from various sites on the Internet or, if you feel inclined, even design your own sheets. It doesn't really matter what you use, or how you record the data, so long as you include the basic information such as object name, type, position, constellation and size. But you can also add more personal information such as magnification, eyepieces or filters used, and the transparency and seeing conditions of your observing site.

Digital imagers who record their observations onto CCD or other electronic media should not feel excluded but if paper-based logging does not appeal; the digital files can at least be coded in the filename, or better still, imprinted onto the actual image itself. Otherwise you will end up with masses of digital images, and no idea of when or how they were created. File naming structures could include the following data; object name, date and exposure, for example "M13_24/10/05_3×10_Secs.tiff".

Digital observers should still use a logbook of some sort to list all their completed observations, and not just rely on a hard disk full of electronic files. Organizing and structuring digital data is just as important as filing analogue reports, but if you still cannot face a paper trail of information, there are a number of software packages that can help. Observation logging is integrated into many planetarium programs, which I have mentioned briefly in the chapter on planning and resources, and the standalone packages in that chapter can also help with recording observations as well. Alternatively, if you have access to a database or spreadsheet program such as Filemaker or Excel, you can easily create your own customized logs.

Whatever imaging or recording format you use, make any notes informative and descriptive, comments such as "very impressive" will not mean an awful lot if you look back at this information a year later. Describe what you see, and record this accurately, even if the meaning has no technical or scientific value. If the cluster looks like sprinkled sugar, or appears in the shape of a Christmas tree (one cluster actually does), then say so. On your travels of the night sky, also make a note of other interesting objects that you can follow up on future observing sessions.

From a personal point of view, I would encourage observers at all levels to share their work with local societies, and send your reports in to the national societies such as the SPA, BAA and the Webb Society, who will welcome these observations with open arms.

Chapter 15

Comprehensive Observing List

Creating a list of target observations, even under the specific heading of star clusters, is a difficult process, especially within the limits imposed by a book of this size. Deciding what to include and which objects to omit is largely down to personal preference. I have, therefore, tried to include a range of objects from visually easy to very difficult clusters, and a selection somewhere in between. Hopefully, this will give observers in all stages of development – whether beginner, intermediate or advanced – a range of objects to discover and experience.

Telescope aperture has a significant bearing on the ability to clearly observe these clusters; however, a dedicated and experienced observer can often see more in a small telescope, than an observer with a large telescope who has not yet developed their observing skills. My own personal observations of these clusters were performed with an 8" f6 Newtonian, a 4" f5 Rich-Field Refractor, and a pair of 8 × 50 binoculars. A selection of eyepieces from 6mm to 40mm and a 2× Barlow were also used. The visual descriptions are based on typical suburban skies, for an observer with general eyesight. The observation notes have purposely been kept brief for two reasons; first, due to limitations of space, and second, to allow you to make your own personal "discoveries," but they should be sufficient to locate and confirm each object.

As mentioned previously, field orientation can be confusing at the best of times, due to the differences between celestial and Earth coordinates, and the inversion caused by telescope optics. To ease this problem, all finder charts have been labeled with celestial North and East, and the location descriptions follow these criteria. Many of the clusters within this section are displayed in the finder charts, but due to space restrictions there have been some omissions. Stars are plotted down to magnitude 8.5, and the cluster legends follow standard protocol: a dotted circle for an open cluster, and a circle containing a cross for globular clusters.

The objects in these observing lists consist mainly of open and globular clusters, but there is also a sprinkling of asterisms and stellar associations thrown in to add some variety. As well as standard astronomical data such as size and stellar co-ordinates, each object has a visual description and notes on how to find the cluster, along with any relevant information of interest to the visual observer. Most of these objects are also suitable targets for CCD imaging using typical amateur equipment from suburban locations. The Messier and NGC catalogues have been drawn on in particular to produce this list, but you will also

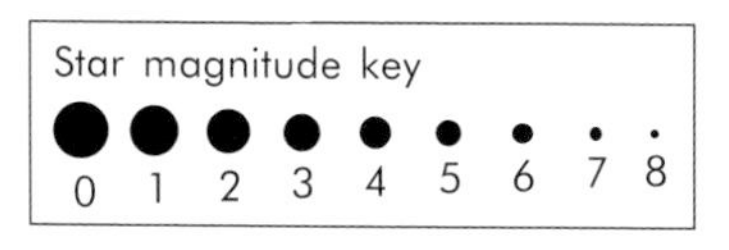

Figure 15.1. Star magnitude key for finder charts.

find objects from more obscure catalogues that have no NGC number. To aid the planning and observation of these clusters, the objects are listed in order of constellation in terms of their Right Ascension. This means that on any given observing night, once you have ascertained that a particular constellation is visible, and is at an acceptable altitude above the horizon, you can work your way through several objects in the list as the night progresses, simply by turning the page.

Working out your declination limit. In other words, the objects visible above your local horizon can be achieved simply by subtracting your latitude from 90°. For example, in the UK, the latitude is approximately 51° 30' North, and subtracting this from 90° gives a declination limit of –38.5°. However, a more realistic figure, taking into account the usual horizon obstacles of suburbia is about –30° in the UK. Remember that the declination limit is negative only because the objects lie below the "celestial" equator. Similarly, from New York, the latitude is 40° 42' N, so the actual declination limit is –49.5°, but in reality observing objects down to about –43° is more likely to be achieved.

Within this observing list, most of the clusters are detectable from mid-northern latitudes, but there are several objects that can only be seen from a more southerly locale.

All the clusters in this chapter have also been assigned with a "difficulty" rating to aid in selecting objects appropriate to your equipment and skill levels, however do not be afraid to select a "medium" or "hard" object just because you have a small telescope. A large number of these clusters can be seen in most instruments, but in smaller ones they will appear fainter and be more difficult to resolve. You may also notice that some bright objects have been given a medium or hard rating because they are more difficult to find due to barren star fields or nondescript constellations. The "difficulty" rating is broken down as follows, but is a guide only, and there will certainly be some overlap between categories and the clusters within them.

Easy

This category includes clusters especially suited to beginners because they are generally larger and brighter than their counterparts. As a result, they should also be easier to locate and to resolve any detail. Experienced observers and large telescope users should not be put off though, as this category contains many of the finest star cluster examples. Some of the clusters in this class can be seen with binoculars, but a small telescope will give a better view, and from very dark sites you may be able to spot the odd one with the naked eye. Many of the Messier and bright NGC objects fall into this category.

Medium

Objects listed as "medium" difficulty are normally fainter and smaller than those listed as "easy" above and may be harder to locate. A small to medium telescope from 4" is required to observe these clusters in any detail, and binoculars are not generally suitable. Using a larger telescope will improve the visual resolution and allow more detail to be confirmed. In this category, many NGC clusters have been listed.

Hard

Most of the objects with this designation are best observed with a medium or large telescope 8" or even larger. These clusters will often be fainter, less dense and more difficult to resolve than easy or medium targets. Larger telescopes, with apertures of 10" or more will enable more structure and detail to be extracted than smaller models. The fainter NGC objects and several more obscure clusters lie in this category.

Difficult

Last but not least, the "difficult" clusters range from those visible in large telescopes, 10" to 12" in diameter, and to the extreme objects requiring very large telescopes 16"- to 20" or more to observe with any clarity. A few of these objects are CCD or large telescope objects only, but some can be seen using averted vision with smaller scopes, patience and perseverance. Difficult objects include faint NGC and IC objects, very faint open clusters and several distant or extragalactic globulars.

• NGC 104

Other Name	Type	Const	Mag	Class
47 Tucanae	Globular	Tucana	3.95	III
RA	**Dec**	**Size**	**Distance**	**Rating**
00h 24m 05.2s	–72° 04' 51"	50'	14,700 Lyr	Easy

Description and Notes

As NGC 104 is the second brightest globular in the night sky, it deserves a place in any cluster catalogue. It is similar in size and appearance to Omega Centauri, the brightest globular, and is well resolved with modest optical aid, such as a 4" telescope, and can easily be seen with binoculars. Some observers note a pale yellow hue, especially towards the core, which is highly concentrated. NGC 104 is situated about 5° from the magnitude 2.7 star beta Hydri, is close to the SMC, and about 2° from another globular NGC 362 (Caldwell 104) at magnitude 6.4.

• G1

Other Name	Type	Const	Mag	Class
Mayall II	Globular	Andromeda	13.7	–
RA	**Dec**	**Size**	**Distance**	**Rating**
00h 32m 46s	+39°34'41"	–	2,900 Kly	Difficult

Figure 15.2. G1 an extragalactic globular cluster © Tony O'Sullivan.

Description and Notes

M31, the Andromeda galaxy contains over 500 globular clusters; so G1, the brightest member, is actually an extragalactic globular – it does not reside in the Milky Way. This cluster is extremely faint but can be picked up visually in large telescopes 12 to 16" where a diffused glow is apparent showing a slightly stellar structure. G1 is also a good target for CCD imaging and is located roughly 2.5° South West of the galaxy's center.

• Hodge 5

Other Name	Type	Const	Mag	Class
–	Globular	Cassiopeia	16.7	–
RA	**Dec**	**Size**	**Distance**	**Rating**
00h 39m 14s	+48°23'06"	–	–	Difficult

Description and Notes

Hodge 5 is another extragalactic globular cluster that resides within NGC 185, a magnitude 9 elliptical galaxy in Cassiopeia. At magnitude 16.7, this cluster is extremely faint and should not even be attempted without a massive telescope in the 20" region and a detailed finder chart. This object is located about 3.5° North East of the galactic nucleus and even with a "light bucket" it appears stellar with averted vision, showing no detail or structure. Hodge 5 is by far the most difficult object within this observing list.

• Cas OB 1

Type	Const	RA	Dec	Rating
Association	Cassiopeia	01h 00m 8s	+61°30'00"	Easy

Description and Notes

Centered around the bright star gamma Cassiopeiae, this stellar association is an easy target in binoculars or 3 to 4" wide-field telescopes, where over two dozen stars can be detected between magnitudes 6 to 10, in a field over 1° wide. There are six such associations in Cassiopeia alone, which are perfect for small telescope users.

• NGC 457

Other Name	Type	Const	Mag	Class
Caldwell 13	Open	Cassiopeia	6.4	I 3 r
RA	**Dec**	**Size**	**Distance**	**Rating**
01h 19m 35s	+58°17'12"	20'	7,918 Lyr	Easy

Description and Notes

NGC 457 has been nicknamed the "Owl Cluster" and it certainly lives up to this moniker, with its "eyes" peering at you. Even the celestial outstretched "wings" are prominent in my 4" refractor at 50×. Observing this cluster with an 8" telescope, I detected many more stellar members and conclude it is probably the finest example in this constellation, located 2° South West of delta Cassiopieae.

• M103

Other Name	Type	Const	Mag	Class
NGC 581	Open	Cassiopeia	7.4	III 2 p
RA	**Dec**	**Size**	**Distance**	**Rating**
01h 33m 23s	+60°39'00"	5'	7,152 Lyr	Medium

Description And Notes

Using an 8" Newtonian, I originally mistook this object for NGC 457 which is 3° away. However, this cluster is much smaller and dimmer, containing over 40 stars from magnitude 10. Situated slightly more than 1° North East of delta Cassiopieae, the main stars form a triangular shape, vaguely like a Christmas tree. M103 is fairly compact and is just visible in binoculars.

Figure 15.3. NGC 457 © Tony O'Sullivan.

• C39

Other Name	Type	Const	Mag	Class
–	Globular	Triangulum	15.9	–
RA	**Dec**	**Size**	**Distance**	**Rating**
01h 34m 50s	+30°21'56"	–	–	Difficult

Description and Notes

Not to be confused with Caldwell 39 which is a planetary nebula in Gemini; C39 in this case is an extragalactic globular cluster associated with M33, the Triangulum galaxy. As with most extragalactic clusters, this object is very faint but has been spotted in telescopes of 12" aperture, although a 16" model will show a non-stellar core. C39 is certainly challenging, but CCD users in particular may wish to seek C39 out, which is situated about 20' South East of the galaxies center.

• NGC 752

Other Name	Type	Const	Mag	Class
Caldwell 28	Open	Andromeda	5.7	III 1 m
RA	**Dec**	**Size**	**Distance**	**Rating**
01h 57m 41s	+37°47'06"	75'	1,490 Lyr	Easy

Figure 15.4. NGC 752 © Tony O'Sullivan.

Description and Notes

A large, very open cluster that requires a wide field of view to encompass; hence the use of my 4" f5 refractor, where I observed several double stars and over 30 stellar members. NGC 752 actually contains over 80 stars and is located 3.5° North West of beta Trianguli. It is a great binocular object if skies permit, and shows differences in star colors under close scrutiny.

• NGC 884 / 869

Other Name	Type	Const	Mag	Class
"h" & "chi" Persei	Open	Perseus	6.1 / 5.3	I 3 r (both)
RA	**Dec**	**Size**	**Distance**	**Rating**
02h 22m 18s	+57°08'12"	18'	7,644 Lyr	Easy
02h 19m 00s	+57°07'42"	18'	6,777 Lyr	Easy

Description and Notes

Although NGC 884 and 869 are completely separate clusters, due to their close proximity of less than half a degree they are generally observed together and commonly known as the "Double Cluster." Visible with the help of binoculars, the pair make a mesmerising sight and a small wide field telescope at low power can resolve both clusters in a 1° field. The clusters are similar in size and structure, though 884 is slightly brighter and more condensed. Both clusters also look great

Figure 15.5. M34 © Tony O'Sullivan.

in larger telescopes especially if you can view them together, and in my 8" f6, I resolved many stellar members at 30×. To find the duo, locate eta Persei and traverse 4.5° North West.

• M34

Other Name	Type	Const	Mag	Class
NGC 1039	Open	Perseus	5.2	II 3 m
RA	**Dec**	**Size**	**Distance**	**Rating**
02h 42m 05s	+42°45'42"	35'	1,500 Lyr	Easy

Description and Notes

Clearly visible in binoculars, M34 is an impressive bright cluster that is also well suited to rich-field refractors due to its large diameter. Many of the 100 member stars can be viewed in a 4" scope especially near the cluster's center. A larger scope does not really improve the view, but more stars will be resolved. M34 almost filled the eyepiece at 48× in my 8" f6, and I observed a "square" stellar core and several curving arcs of stars situated 5° North West of beta Persei (Algol).

• Collinder 34

Other Name	Type	Const	Mag	Class
–	Open	Cassiopeia	6.8	I 3 p
RA	**Dec**	**Size**	**Distance**	**Rating**
02h 59m 23s	+60°34'00"	25'	–	Medium

Figure 15.6. NGC 1245 © Tony O'Sullivan.

Description and Notes

Collinder 34 is another large but fairly nondescript cluster in Cassiopeia, and is quite faint with only a couple of stars brighter than magnitude 8 or 9. This cluster also lies within a large nebula, IC1848. With my 4" refractor, I detected about 20 stars in a low power eyepiece at 50×. To find this object, move your telescope roughly 13° North of eta Persei and look for a swath of faint stars.

• Trumpler 3

Other Name	Type	Const	Mag	Class
Harvard 1	Open	Cassiopeia	7.0	III 3 p
RA	**Dec**	**Size**	**Distance**	**Rating**
03h 11m 48s	+63°15'00"	23'	2,715 Lyr	Medium

Description and Notes

This cluster is quite close to Stock 23, but due to its larger size, I actually found it easier to spot, even though it is not as bright. Many stars can be seen in a 4" (I detected 22 at 100×) but they are quite faint, at magnitude 9 or less. The cluster is fairly loose and irregular in shape, and is located 9° North East of gamma Persei.

Figure 15.7. NGC 1342 © Tony O'Sullivan.

• NGC 1245

Other Name	Type	Const	Mag	Class
OCL 389	Open	Perseus	8.4	III 1 r
RA	**Dec**	**Size**	**Distance**	**Rating**
03h 14m 42s	+47°14'12"	40'	9,128 Lyr	Medium

Description and Notes

A small, fairly dim, nondescript cluster, positioned roughly between alpha and kappa Persei. Using my 4" refractor, only about 14 cluster stars were visible, although this object contains over 200 stars from magnitude 11. A larger 12" will reveal 70 or more densely packed stars and a more uniform structure. To completely resolve NGC 1245 a 16" instrument is required, which reveals over 100 stars arranged in lanes.

• Stock 23

Other Name	Type	Const	Mag	Class
–	Open	Camelopardalis	5.6	III 3 p n
RA	**Dec**	**Size**	**Distance**	**Rating**
03h 16m 11s	+60°06'56"	14'	3,260 Lyr	Medium

Description and Notes

Stock 23 can be seen in binoculars as a faint group of stars, but under less than perfect skies a 4" will give better views. With my small refractor I managed to resolve 18 stars at 100×, with four bright members forming a diamond shape though there are, in fact, over 25 stars. This cluster also has associated nebulosity, but I could not detect it. Stock 23 is just over 6° North East of gamma Persei.

• NGC 1342

Other Name	Type	Const	Mag	Class
OCL 401	Open	Perseus	6.7	III 3 p
RA	**Dec**	**Size**	**Distance**	**Rating**
03h 31m 38s	+37°22'36"	15'	2,168 Lyr	Medium

Description and Notes

Using an 8" telescope a large, bright, fairly loose cluster can be observed which will uncover about 20 or so stars. A very large telescope in the region of 16" will resolve at least 60–70 members that appear as chains and strings of stars. The most luminous stars in this cluster are magnitude 8, and the group is located 5.5° South West of epsilon Persei.

Figure 15.8. Palomar 1, a very faint globular © Tony O'Sullivan.

• Palomar 1

Other Name	Type	Const	Mag	Class
Pal 1	Globular	Cepheus	13.2	XII
RA	**Dec**	**Size**	**Distance**	**Rating**
03h 33m 23s	+79°34'50"	2.8'	35,600 Lyr	Difficult

Description and Notes

Palomar 1 is one of the faintest globular clusters in the night sky and should not be tackled without a large telescope or CCD equipment. In fact, a 16" telescope only reveals the cluster as a faint sphere with averted vision. A large observatory class instrument, 20" or more can detect the cluster more readily, but does not reveal much detail or resolution. However, using a quality CCD unit this globular can be imaged using a 10" telescope. If you wish to take on the challenge, look about 9.5° North of Polaris for an extremely faint glow.

• M45

Other Name	Type	Const	Mag	Class
Melotte 22	Open	Taurus	1.5	I 3 r n
RA	**Dec**	**Size**	**Distance**	**Rating**
03h 47m 00s	+24°07'00"	120'	450 Lyr	Easy

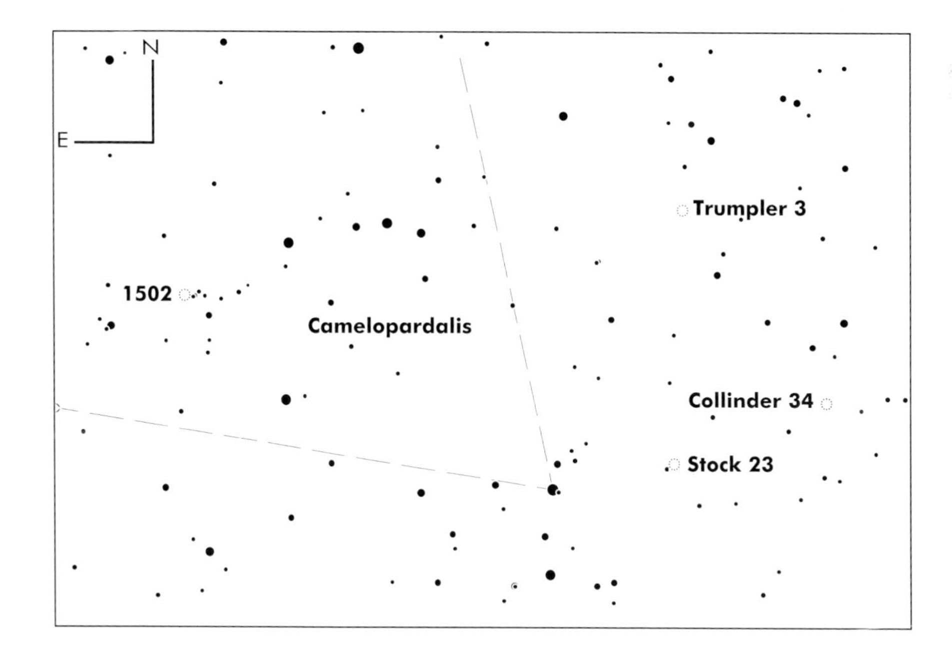

Figure 15.9. Finder Chart for Camelopardalis © Software Bisque.

Figure 15.10. NGC 1513 © Tony O'Sullivan.

Description and Notes

Visible as a naked eye object practically anywhere above the horizon, M45 or the "Pleiades" is one of the brightest open clusters and is best observed in binoculars or small telescopes to appreciate its beauty. The stellar members shine with an intense blue color, reminiscent of a "mini plough" like that of Ursa Major. To examine the nebulosity associated with this cluster, I used my 8" at 133×, but general viewing demands low power fields. You cannot miss this cluster, but it is located 13° North West of alpha Tauri.

• Kembles Cascade

Other Name	Type	Const	Mag	Class
–	Asterism	Camelopardalis	4.0	–
RA	**Dec**	**Size**	**Distance**	**Rating**
03h 57m 45s	+63°04'00"	2.5 deg	–	Easy

Description and Notes

One of the most interesting "mere" asterisms I have come across, Kembles Cascade is an impressive bright chain of stars spanning over 2 degrees, which shows about 13 stars in a 4" telescope. A low power eyepiece provides the best view of this object, which forms an almost straight line towards the open cluster, NGC 1502.

• NGC 1502

Other Name	Type	Const	Mag	Class
OCL 383	Open	Camelopardalis	6.9	II 3 p
RA	**Dec**	**Size**	**Distance**	**Rating**
04h 07m 50s	+62°19'54"	8'	2,676 Lyr	Medium

Description and Notes

A small, fairly compressed open cluster that is quite faint, but easily found by following the chain of stars that form "Kemble's Cascade" from NE to SW, or by locating beta Camelopardalis and moving 7° North West. NGC 1502 forms a triangular shape with two central brighter stars, and about 25 faint members in my 4" f5 refractor.

• NGC 1513

Other Name	Type	Const	Mag	Class
OCL 398	Open	Perseus	8.4	II 1 m
RA	**Dec**	**Size**	**Distance**	**Rating**
04h 09m 57s	+49°30'54"	10'	4,303 Lyr	Hard

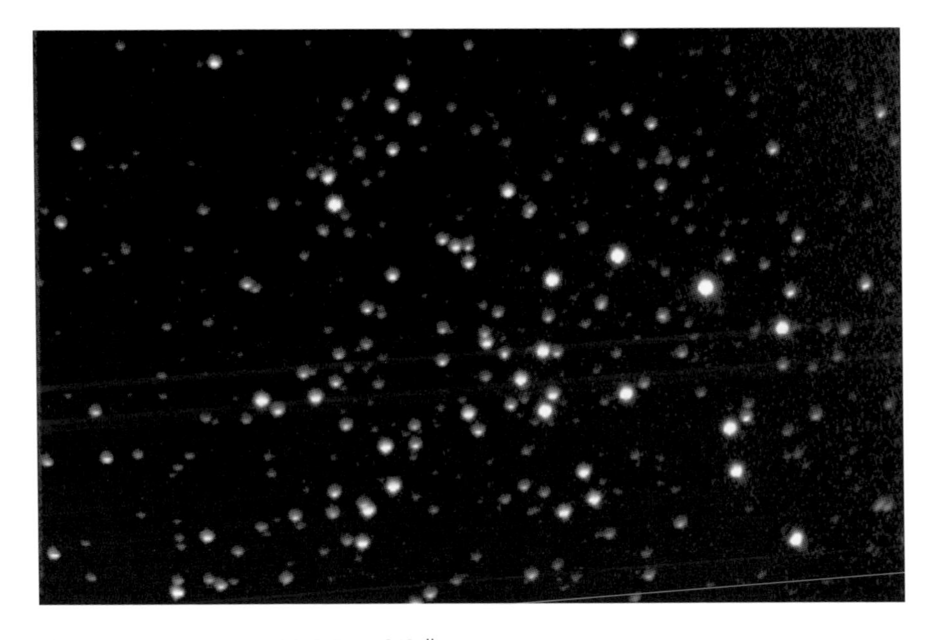

Figure 15.11. NGC 1528 © Tony O'Sullivan.

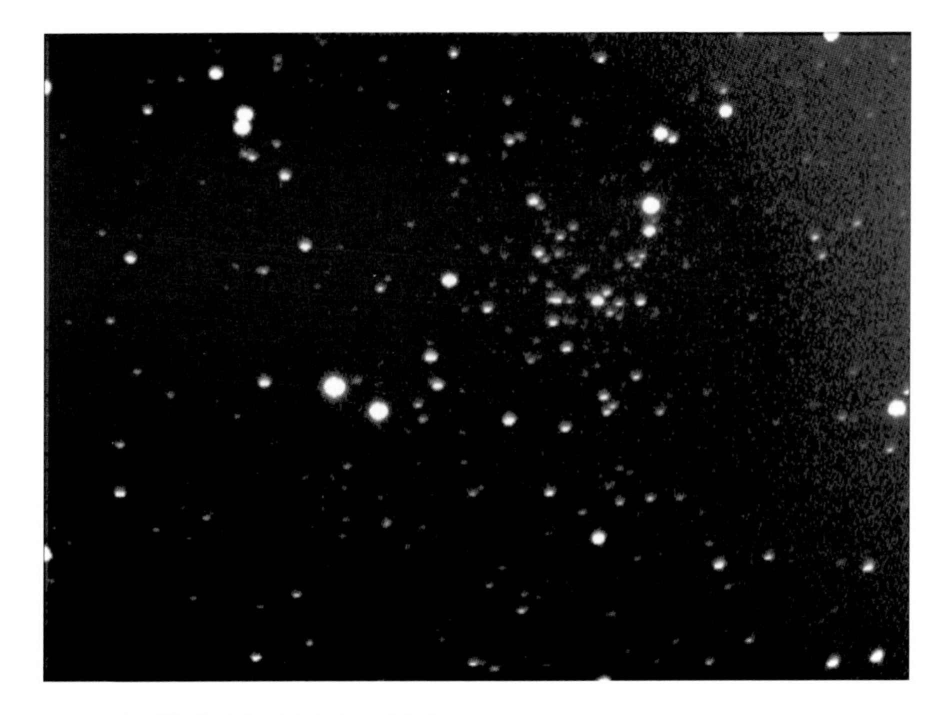

Figure 15.12. Berkeley 11 © Tony O'Sullivan.

Description and Notes

NGC 1513 is located 1° South East of lambda Persei, and is a fairly dense cluster containing about 50 stars with some compression on the Western edge. The brightest is magnitude 11, so medium telescopes are better suited to this object, and will reveal 20–30 member stars. However, users with 12" aperture can resolve most of the component stars and detect many unresolved field stars.

• NGC 1528

Other Name	Type	Const	Mag	Class
OCL 397	Open	Perseus	6.4	II 2 m
RA	**Dec**	**Size**	**Distance**	**Rating**
04h 15m 23s	+51°12'54"	16'	2,529 Lyr	Medium

Description and Notes

NGC 1528 is fairly bright and contains about 40 stars with the brightest member at magnitude 8. Small telescopes show it well, about 1.5° North East of lambda Persei, where there are several open clusters nearby including NGC 1513 and 1545. A larger telescope 8 to 10" can resolve up to 80 members, which include faint field stars.

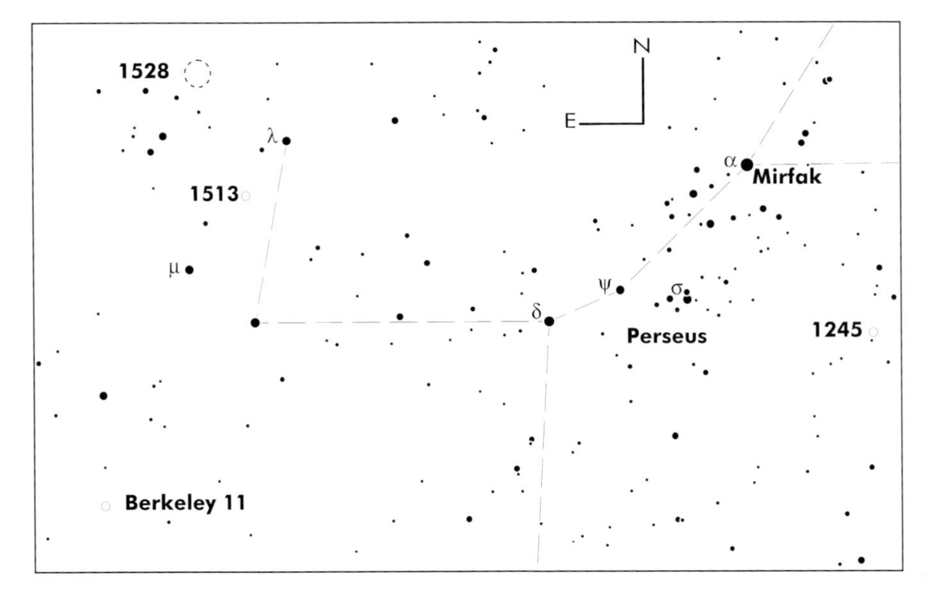

Figure 15.13. Finder Chart 1 for Perseus © Software Bisque.

• Berkeley 11

Other Name	Type	Const	Mag	Class
Berk 11	Open	Perseus	10.4	II 3 m
RA	**Dec**	**Size**	**Distance**	**Rating**
04h 20m 36s	+44°55'00"	5'	7,172 Lyr	Hard

Description and Notes

Berkley 11 is a faint and quite small open cluster in Perseus that contains about 35 stars although the brightest member shines at a feeble magnitude 11.8. Medium or large scopes 8 to 10" are therefore required to obtain a good view. Located 4° South East of 48 Persei, the main stars form a simple arrowhead shape.

• Hyades

Other Name	Type	Const	Mag	Class
Melotte 25	Open	Taurus	0.5	II 3 m
RA	**Dec**	**Size**	**Distance**	**Rating**
04h 27m 00s	+15°57'48"	5.5 deg	147 Lyr	Easy

Description and Notes

The Hyades forms part of the Taurus moving cluster and is situated between the "bull's horns" of the constellation. Being large and bright, the cluster is well presented in binoculars or a rich-field telescope, where over 40 stars can be seen the brightest of which is magnitude 3.4. This is one of the best binocular open clusters.

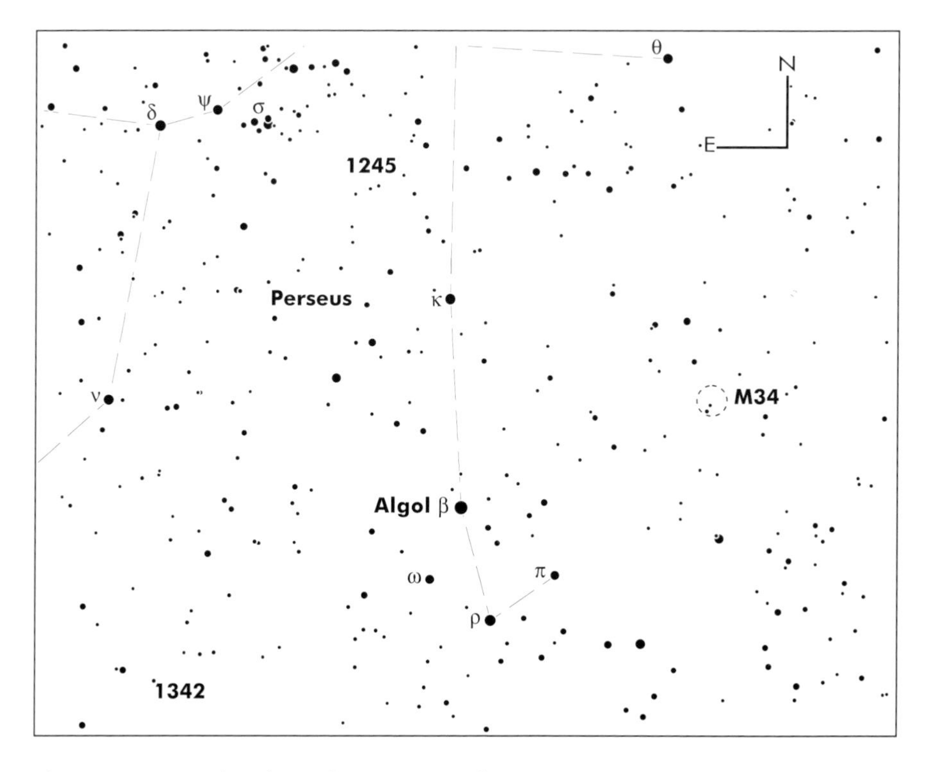

Figure 15.14. Finder Chart 2 for Perseus © Software Bisque.

• Palomar 2

Other Name	Type	Const	Mag	Class
Pal 2	Globular	Auriga	13	IX
RA	**Dec**	**Size**	**Distance**	**Rating**
04h 46m 06s	+31°22'51"	2.2	90,000 Lyr	Difficult

Description and Notes

Palomar 2 is another difficult globular to detect and observe, and is not for the faint-hearted (no pun intended!). Situated approximately 3° South West of iota Auriga, this object is challenging at magnitude 13 and requires a large aperture 16" plus, and plenty of magnification. Do not expect to resolve any stars or reveal any detail, though CCD imaging may pick up some slight mottling.

• NGC 1817

Other Name	Type	Const	Mag	Class
OCL 463	Open	Taurus	7.7	III 1 m
RA	**Dec**	**Size**	**Distance**	**Rating**
05h 12m 15s	+16°41'24"	16'	6,428 Lyr	Medium

Figure 15.15. Palomar 2 © Digitized Sky Survey.

Description and Notes

NGC 1817 shares a 30' field of view with another open cluster NGC 1807, and although NGC 1817 is fainter, it has a more interesting shape and structure. Using my 4" refractor I noted this object was tightly packed with faint stars, some of which I could only just detect. An 8" will render about 60 stars in a fairly rich group about 15' in diameter. Situated 7.5° South West of zeta Tauri, NGC 1817 and 1807 have been dubbed the "poor man's" double cluster.

• NGC 1893

Other Name	Type	Const	Mag	Class
OCL 439	Open	Auriga	7.5	II 2 m n
RA	**Dec**	**Size**	**Distance**	**Rating**
05h 22m 44s	+33°24'42"	25'	19,560 Lyr	Medium

Figure 15.16. NGC 1817 © Cliff Meredith.

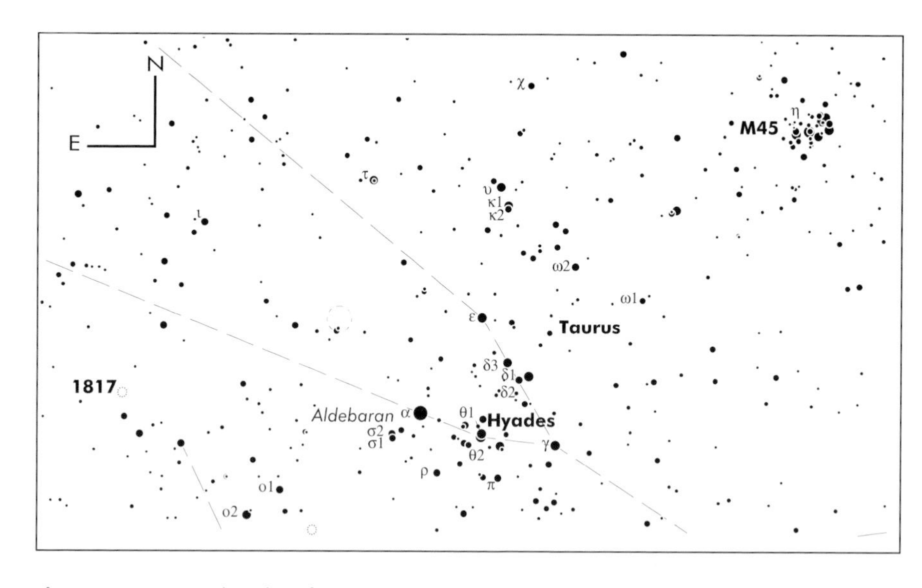

Figure 15.17. Finder Chart for Taurus © Software Bisque.

Figure 15.18. NGC 1893 © Tony O'Sullivan.

Figure 15.19. M38 © Cliff Meredith.

Figure 15.20. M36 © Tony O'Sullivan.

Description and Notes

This cluster is not as impressive visually as nearby M36 or M38 in Auriga, but it is a pleasant, fairly tight cluster associated with the flaming star nebula, IC 410. In my 4" telescope, I observed several tight groupings of stars, and many faint members but did not detect the nebula. Located 5° East of iota Aurigae, the cluster is easy to find and is part of the Auriga OB association.

• Berkeley 19

Other Name	Type	Const	Mag	Class
–	Open	Auriga	11.4	IV 2 p
RA	**Dec**	**Size**	**Distance**	**Rating**
05h 24m 06s	+29°36'00"	4'	15,749 Lyr	Difficult

Description and Notes

Berkeley 19 is a very faint cluster in Auriga that is really best left for large telescopes and digital imagers. With over 40 stars, you may assume it would be easy to detect, but the brightest member is a mere magnitude 14.7. The cluster is situated just over 1° North West of beta Tauri if you wish to give it a try.

• M38

Other Name	Type	Const	Mag	Class
NGC 1912	Open	Auriga	6.4	III 2 m
RA	**Dec**	**Size**	**Distance**	**Rating**
05h 28m 00s	+35°51'02"	15'	4,200 Lyr	Medium

Description and Notes

This cluster is only 2° from the brighter cluster M36, also in Auriga, and in my 8" f6 this cluster appeared larger, but contained many faint stars that required high power to resolve. M38 is evenly distributed and at 133× resembled a crooked cross shape, 6° South West from theta Aurigae.

• Ori OB 1

Type	Const	RA	Dec	Rating
Association	Orion	05h 31m 4s	–02°41'00"	Easy

Description and Notes

Ori OB1 is a stellar association that is actually centered around the region of M42 the infamous Orion Nebula. Within this stellar grouping are the Trapezium,

Figure 15.21. M37 © Tony O'Sullivan.

Collinder 70 and NGC 1981 open clusters. Around 30 stars can be detected in a 3 or 4" telescope, and this object is also an excellent binocular candidate, but not suitable for large instruments.

• M36

Other Name	Type	Const	Mag	Class
NGC 1960	Open	Auriga	6.0	II 3 m
RA	**Dec**	**Size**	**Distance**	**Rating**
05h 36m 18s	+34°08'24"	10'	4,296 Lyr	Easy

Description and Notes

Situated between the stars theta and iota Aurigae, M36 is one of several bright Messier open clusters in this constellation. A luminous, neat arrangement of stars is presented at low power (48×) in my 8" Newtonian, and the central 4 or 5 stars are particularly prominent. Over 60 stars are associated with this cluster, many of which can also be detected in a 4" telescope, and the outer stars form several curved lanes.

• M37

Other Name	Type	Const	Mag	Class
NGC 2099	Open	Auriga	5.6	II 1r
RA	**Dec**	**Size**	**Distance**	**Rating**
05h 52m 18s	+32°33'12"	14'	4,508 Lyr	Easy

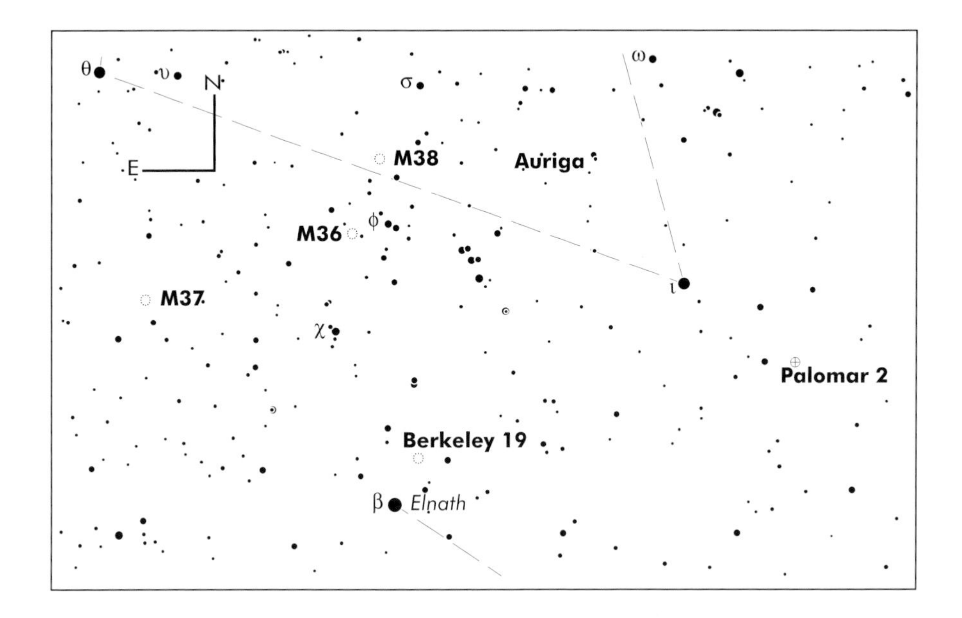

Figure 15.22. Finder Chart for Auriga © Software Bisque.

Figure 15.23. NGC 2129 © Tony O'Sullivan.

Description and Notes

M37 is another relatively bright cluster in Auriga and is particularly impressive, reminiscent of a globular cluster at low power, due to its dense structure. It is located 5° South west of theta Aurigae and provides a fine view in a 4" telescope. Using my 8" f6 at medium power, opened up the cluster and defined the core region, revealing even more of the estimated 500 stellar components.

• NGC 2129

Other Name	Type	Const	Mag	Class
OCL 467	Open	Gemini	6.7	III 3 p
RA	**Dec**	**Size**	**Distance**	**Rating**
06h 01m 07s	+23°19'20"	5'	492 Lyr	Medium

Description and Notes

A fairly small, faint cluster, situated about 42' West of the magnitude 4 star, 1 Geminorum and 2° South West of M35. The central three stars were much brighter in my 4" scope at 100×, forming a neat triangle, and I detected many background stars.

Figure 15.24. IC 2157 © Tony O'Sullivan.

Figure 15.25. M35 © Tony O' Sullivan.

• IC 2157

Other Name	Type	Const	Mag	Class
–	Open	Gemini	8.4	III 2 p
RA	**Dec**	**Size**	**Distance**	**Rating**
06h 04m 50s	+24°03'21"	5'	6,650 Lyr	Hard

Description and Notes

45' North of 1 Geminorum, lies the open cluster IC 2157 in Gemini, which is one of over a dozen in this constellation, containing about 20 stars with the brightest member at magnitude 11. Visually it is faint, so a medium sized telescope is your best bet, to tease-out the less luminous stars.

• M35

Other Name	Type	Const	Mag	Class
NGC 2168	Open	Gemini	5.1	III 2 m
RA	**Dec**	**Size**	**Distance**	**Rating**
06h 09m 00s	+24°21'00"	25'	2,973 Lyr	Easy

Figure 15.26. NGC 2192 © Tony O'Sullivan.

Figure 15.27. NGC 2244 © Tony O'Sullivan.

Description and Notes

Although bright, M35 is a subtle open cluster that can be glimpsed in binoculars as a faint hint of stars, though with a 4" telescope the cluster really starts to shine. Positioned just over 2° North West of the magnitude 3 star, eta Geminorum, several faint interconnecting chains of stars can be observed and the clusters overall structure is pretty loose. With an 8" instrument at 133×, I detected about 40 or so member stars but M35 actually contains over 200! Using high power, the central region was particularly appealing at 266×.

• Gem OB 1

Type	Const	RA	Dec	Rating
Association	Gemini	06h 09m 8s	+21°35'00"	Easy

Description and Notes

This wide-field, loose group of stars is a stellar association centered around the stars mu and eta Geminorum and is seen at its best through binoculars or a rich-field telescope. Covering an area approximately 1.5 degrees, it contains a few dozen stars around magnitude 7 to 9, with the open cluster M35 just over 2° to the North West, it is a pleasing sight at low power.

• NGC 2192

Other Name	Type	Const	Mag	Class
OCL 437	Open	Auriga	10.9	III 1 p
RA	**Dec**	**Size**	**Distance**	**Rating**
06h 15m 17s	+39°51'18"	5'	11,410 Lyr	Difficult

Figure 15.28. NGC 2266 © Tony O'Sullivan.

Description and Notes

NGC 2192 is a small, very faint cluster located just under 6° North West of alpha Aurigae, and even with an 8" telescope it is difficult to resolve any stellar members.

A 12 to 14" instrument improves matters, but as the brightest stars are a mere magnitude 14, even this size of aperture only reveals 10 or so faint stars embedded in a diffused haze. There are 45 stars associated with this cluster, and CCD imagers will fare much better.

• NGC 2244

Other Name	Type	Const	Mag	Class
Caldwell 50	Open	Monoceros	4.8	II 3 r n
RA	**Dec**	**Size**	**Distance**	**Rating**
06h 31m 55s	+04°56'30"	29'	4,710 Lyr	Easy

Description and Notes

This open cluster is associated with the Rosette nebula, a molecular cloud over 1° in diameter, with NGC 2244 sat in the middle. The cluster is visible in binoculars, and in my 8" telescope I detected over 20 members that had a distinct blue tinge,

Figure 15.29. NGC 2304 © Tony O'Sullivan.

although there are over 100 true component stars. To find the cluster, locate Betelguese (alpha Orionis) and move 9° South East.

• NGC 2266

Other Name	Type	Const	Mag	Class
OCL 471	Open	Gemini	9.5	II 2 m
RA	**Dec**	**Size**	**Distance**	**Rating**
06h 43m 19s	+26°58'12"	5'	11,804 Lyr	Hard

Description and Notes

Slightly less than 7° North of epsilon Geminorum lies the faint open cluster NGC 2266 which is dense, but small and contains about 50 stars between magnitude 11 and 15. A medium telescope in the region of 8" is really required to view this cluster where about 15 or so member stars are portrayed. Larger apertures 12" or more show about 20–30 faint stars.

• NGC 2304

Other Name	Type	Const	Mag	Class
OCL 484	Open	Gemini	10	II 1 p
RA	**Dec**	**Size**	**Distance**	**Rating**
06h 55m 11s	+17°59'18"	3'	13,010 Lyr	Hard

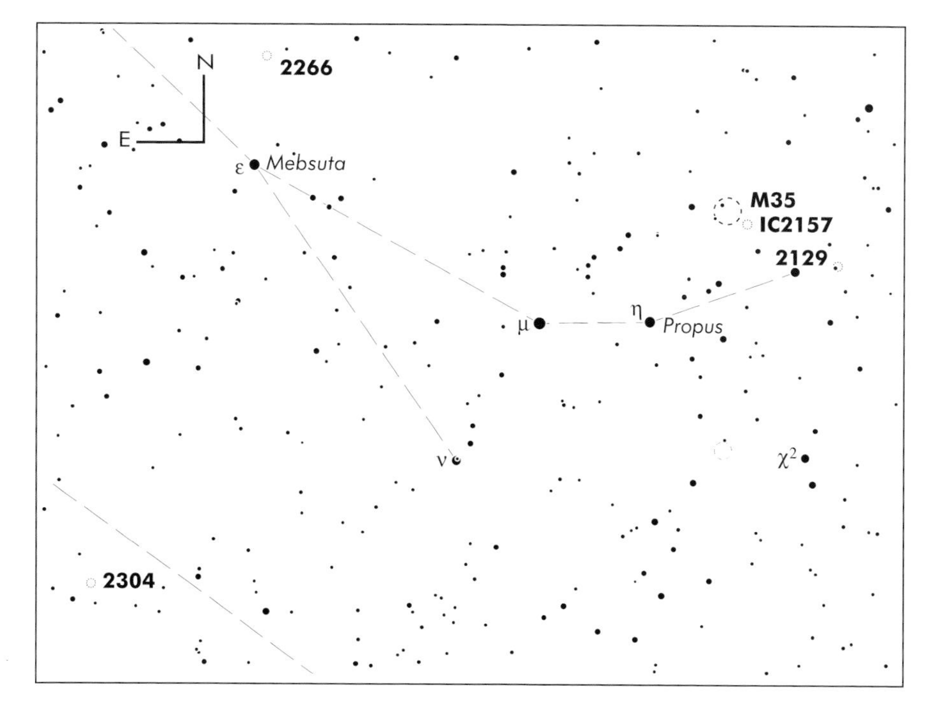

Figure 15.30. Finder Chart for Gemini © Software Bisque.

Description and Notes

This open cluster in Gemini is tiny, at 3' and with a magnitude of 10, it really lends itself to large telescope users, and CCD imagers. Even a 12" aperture at high power only reveals about 20 faint stars in a diffused wedge. The cluster actually contains over 30 stars and is positioned approximately 4.5° North East of gamma Geminorum.

• M50

Other Name	Type	Const	Mag	Class
NGC 2323	Open	Monoceros	5.9	II 3 m
RA	**Dec**	**Size**	**Distance**	**Rating**
07h 02m 42s	–08°23'00"	14'	3,260 Lyr	Medium

Description and Notes

The open cluster M50 in Monoceros, contains 100 stars from magnitude 9–14 and is located 9° North East of the bright star, alpha Canis Majoris. At low power in my 8" Newtonian I observed a distinct cross and pointer shape -+->, formed by the brighter stars which were resolved easily at 48×. As the cluster is situated in a rich Milky Way region, it can be tricky to define, but is worth the effort.

Figure 15.31. M50 © Cliff Meredith.

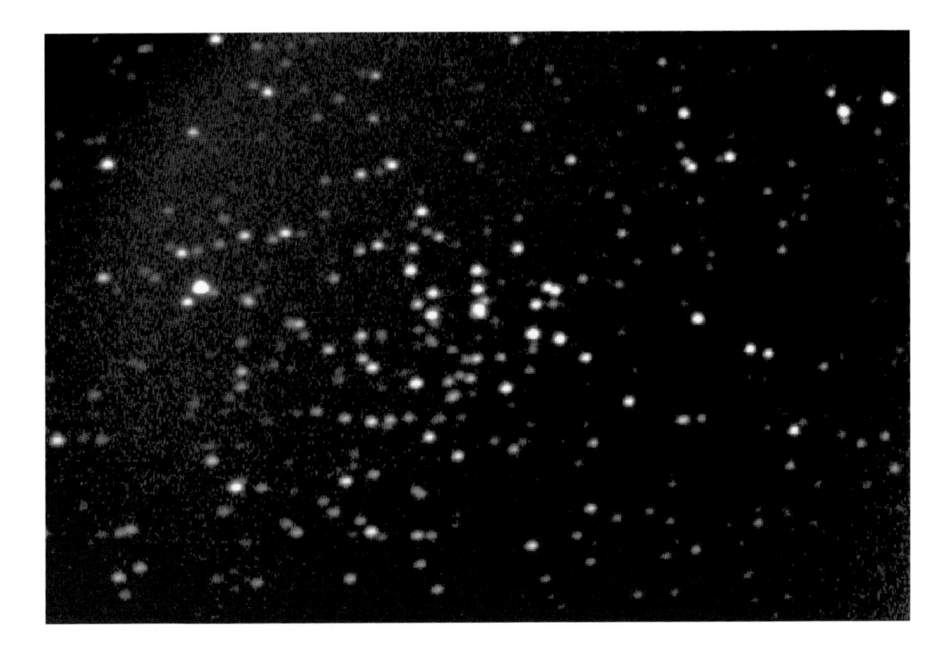

Figure 15.32. NGC 2360 © Tony O'Sullivan.

Figure 15.33. M47 © Tony O'Sullivan.

• NGC 2360

Other Name	Type	Const	Mag	Class
Caldwell 58	Open	Canis Major	7.2	II 2 m
RA	**Dec**	**Size**	**Distance**	**Rating**
07h 17m 43s	–15°38'30"	13'	6,151 Lyr	Medium

Description and Notes

NGC 2360 is a relatively bright cluster containing about 80 stars from magnitude 10.4, and is located approximately 3° East of gamma Canis Majoris. In binoculars, the cluster is detectable as a faint diffused patch, but an 8" telescope resolves the cluster well, showing around 25–30 stars.

• M47

Other Name	Type	Const	Mag	Class
NGC 2422	Open	Puppis	4.4	III 2 m
RA	**Dec**	**Size**	**Distance**	**Rating**
07h 36m 35s	–14°29'00"	25'	1,597 Lyr	Easy

Figure 15.34. NGC 2423 © Tony O'Sullivan.

Description and Notes

M47 is just over 1° North West of M46 another open cluster in Puppis, and contains over 100 stars with the most luminous at magnitude 5. This object contains a varied mix of bright and faint stars including several doubles that can be observed in a 4" telescope, and glimpsed in binoculars. Using my 8" Newtonian, I catalogued around 25 member stars at 133×, though it can be difficult to discern true cluster components from background stars.

• NGC 2423

Other Name	Type	Const	Mag	Class
OCL 592	Open	Puppis	6.7	IV 2 m
RA	**Dec**	**Size**	**Distance**	**Rating**
07h 37m 06s	−13°52'18"	12'	2,497 Lyr	Medium

Description and Notes

Close to M46 and M47, this cluster NGC 2423 contains 40 stellar members with the brightest around magnitude 9. It appeared less structured and more scattered than its more obvious neighbors. At 50× in my 4" I detected several "arms" that appeared to spiral out from its center, about 11° North of rho Puppis.

Figure 15.35. M46 with Planetary Nebula NGC2438 in the field © Cliff Meredith.

• NGC 2419

Other Name	Type	Const	Mag	Class
Caldwell 25	Globular	Lynx	10.4	II
RA	**Dec**	**Size**	**Distance**	**Rating**
07h 38m 8s	+38° 52'55"	4.6'	274,600 Lyr	Hard

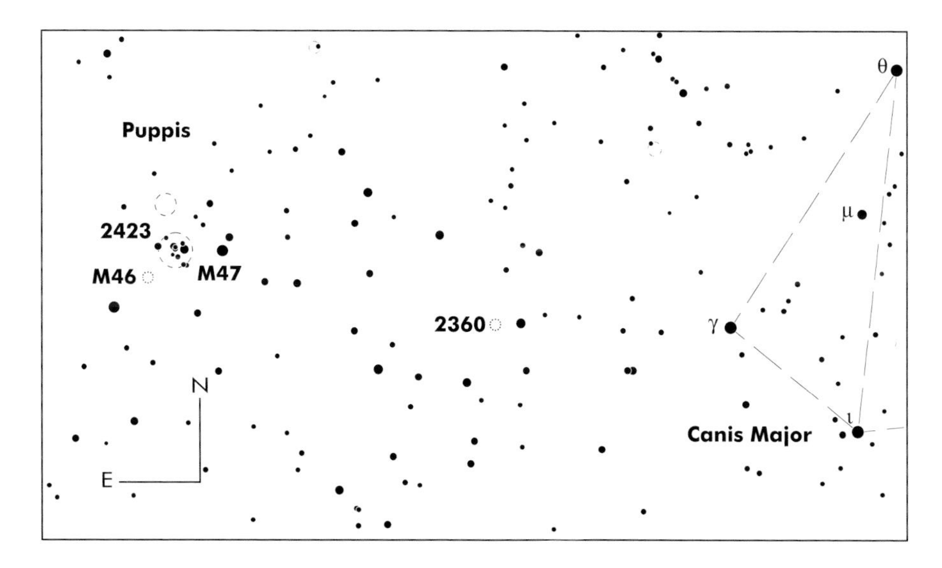

Figure 15.36. Finder Chart for Puppis © Software Bisque.

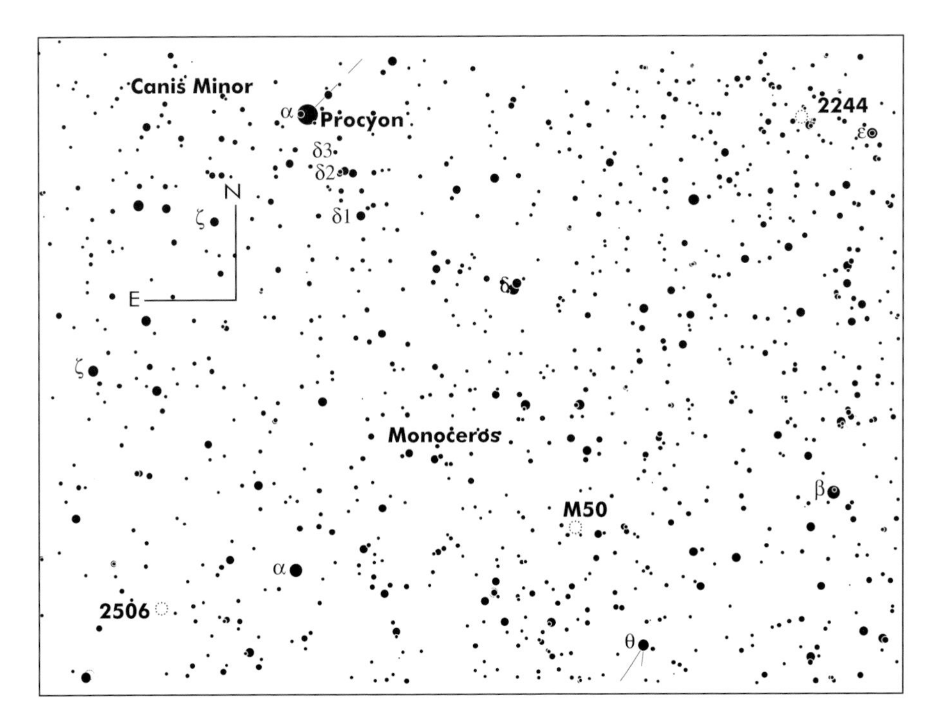

Figure 15.37. Finder Chart for Monoceros © Software Bisque.

Description and Notes

For many years, NGC 2419 was considered the most distant globular cluster in our galaxy, and for this reason alone it is a "must see" object. In my 8" telescope at 50× it appears small and compact, and higher powers of 100×, and even 200× did not reveal any mottling or structure. This object is not really suitable for telescopes under 8" and larger telescopes do not improve the view much. Not impressive visually, but considering its distance, NGC 2419 is well worth a look, and can be found 7° North of alpha Geminorum (Castor).

• M46

Other Name	Type	Const	Mag	Class
NGC 2437	Open	Puppis	6.1	III 2 m
RA	**Dec**	**Size**	**Distance**	**Rating**
07h 41m 46s	–14°48'36"	20'	4,482 Lyr	Medium

Description and Notes

Containing over 100 stellar members, M46 is a large, rich open cluster with a ring of central stars, the brightest at magnitude 8.7 though many are magnitude 10 or fainter. Using 8" optics, the cluster was resolved easily at 48×, and with a medium power of 133×, I counted about 30 component stars. The cluster is fairly symmetrical in shape and can be spotted in smaller scopes, about 9° East of gamma Canis Majoris.

Figure 15.38. M48 © Tony O'Sullivan.

Figure 15.39. M44 © Tony O'Sullivan.

• NGC 2506

Other Name	Type	Const	Mag	Class
Caldwell 54	Open	Monoceros	7.6	I 2 r
RA	**Dec**	**Size**	**Distance**	**Rating**
08h 00m 01s	−10°46'12"	12'	11,279 Lyr	Medium

Description and Notes

This small and tight open cluster was seen well through my 4" refractor, but I could not detect many of the fainter stars, only managing about 24 of the estimated 150 cluster members. From magnitude 10.8, these stars will be much more prominent in an 8" scope and I intend to re-observe this cluster with a larger optic. To find NGC 2506, traverse 13.5° North of rho Puppis.

• M48

Other Name	Type	Const	Mag	Class
NGC 2548	Open	Hydra	5.8	I 2 m
RA	**Dec**	**Size**	**Distance**	**Rating**
08h 13m 43s	−05°45'00"	30'	2,506 Lyr	Easy

Figure 15.40. M67 © Tony O'Sullivan.

Description and Notes

An interesting triangular-shaped open cluster in a very rich star field, M48 is very subtle and delicate. Over 80 stars are associated with the cluster, with the brightest member at magnitude 8.6. Small wide field telescopes, 3 to 5" make the best of this cluster at low power, and I could just make out M48 in 10 × 50 binoculars. At almost the same size as the full Moon, this cluster is large, but merges in with nearby field stars. There are no bright stars close to M48, but the cluster can be found 14° South East of alpha Canis Majoris.

• M44

Other Name	Type	Const	Mag	Class
NGC 2632	Open	Cancer	3.1	II 2 m
RA	**Dec**	**Size**	**Distance**	**Rating**
08h 40m 24s	+19°40'00"	70'	561 Lyr	Easy

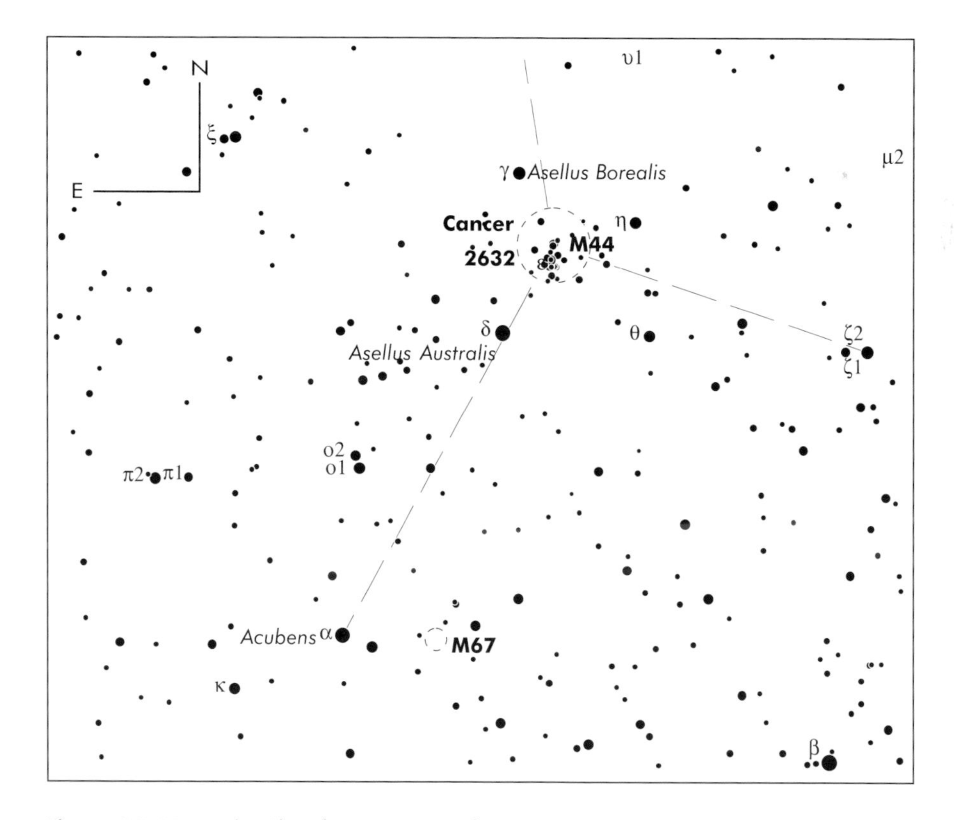

Figure 15.41. Finder Chart for Cancer © Software Bisque.

Figure 15.42. M68 © Tony O'Sullivan.

Description and Notes

Also known as "Praesepe" or the "Beehive Cluster," M44 spans well over a degree in diameter and is a fairly sparse open cluster that contains many double stars. It is a perfect contender for astronomical binoculars or rich-field telescopes, larger optics show more stellar members but perhaps spoil the openness of the view? M44 resides at the intersection point of alpha, iota and zeta Cancri.

• M67

Other Name	Type	Const	Mag	Class
NGC 2682	Open	Cancer	6.9	II 2 m
RA	**Dec**	**Size**	**Distance**	**Rating**
08h 51m 18s	+11°48'00"	25'	2,960 Lyr	Medium

Description and Notes

M67 is visually packed full of tiny stars, and the longer I observe it, more and more stars suddenly pop into view. Stellar counts for this cluster vary from 200 to 500 member stars, but using an 8" at 133×, I managed about 45, and to my mind it is shaped like a "scruffy" number 2! To locate this ancient cluster move 2° South of the magnitude 4 star alpha Cancri.

Figure 15.43. M53 © Tony O'Sullivan.

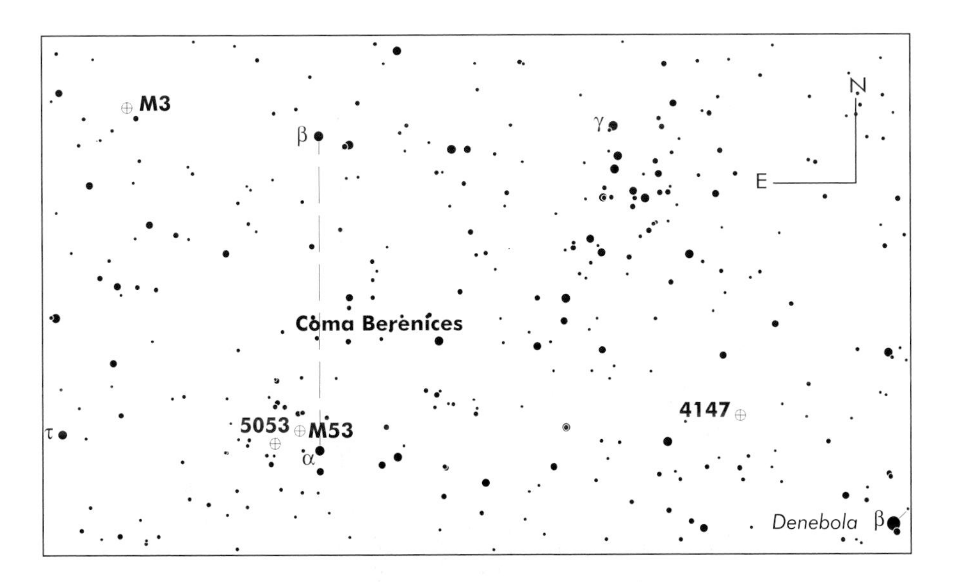

Figure 15.44. Finder Chart for Coma Berenices © Software Bisque.

• NGC 4147

Other Name	Type	Const	Mag	Class
H 1.19	Globular	Coma Berenices	10.32	VI
RA	**Dec**	**Size**	**Distance**	**Rating**
12h 10m 06s	+18°32'31"	4.4'	62,900 Lyr	Hard

Description and Notes

NGC 4147 is the least luminous cluster of the three globulars in Coma Berenices, and is located 6.5° North West of beta Leonis. Only large telescopes will do this globular justice, though an 8" will render it visible but cannot resolve the cluster. Some faint stars and a small nucleus can be resolved using big apertures.

• M68

Other Name	Type	Const	Mag	Class
NGC 4590	Globular	Hydra	7.9	X
RA	**Dec**	**Size**	**Distance**	**Rating**
12h 39m 28s	–26°44'34"	11'	33,300 Lyr	Medium

Description and Notes

To obtain a decent view of M68, a southern location is desirable, but this cluster is just on the limit for UK observers. Small telescopes and possibly bin1oculars should be able to detect this loose, mottled cluster, but an 8" scope will certainly reveal more structure and evidence of several dark lanes. The cluster can be located just over 3.5° South East from the magnitude 2.5 star beta Corvi.

• M53

Other Name	Type	Const	Mag	Class
NGC 5024	Globular	Coma Berenices	7.6	V
RA	**Dec**	**Size**	**Distance**	**Rating**
13h 12m 55s	+18° 10' 09"	13'	58,000 Lyr	Medium

Description and Notes

Visually M53, situated just under a degree from alpha Comae Berenices is difficult to detect in small telescopes and binoculars, and is also situated in a somewhat desolate star field. However, navigating from Arcturus in a straight line, through eta Bootis for approximately 15° should bring you close to the target. With an 8" scope, M53 appeared as a tight unresolved glowing ball at 133×.

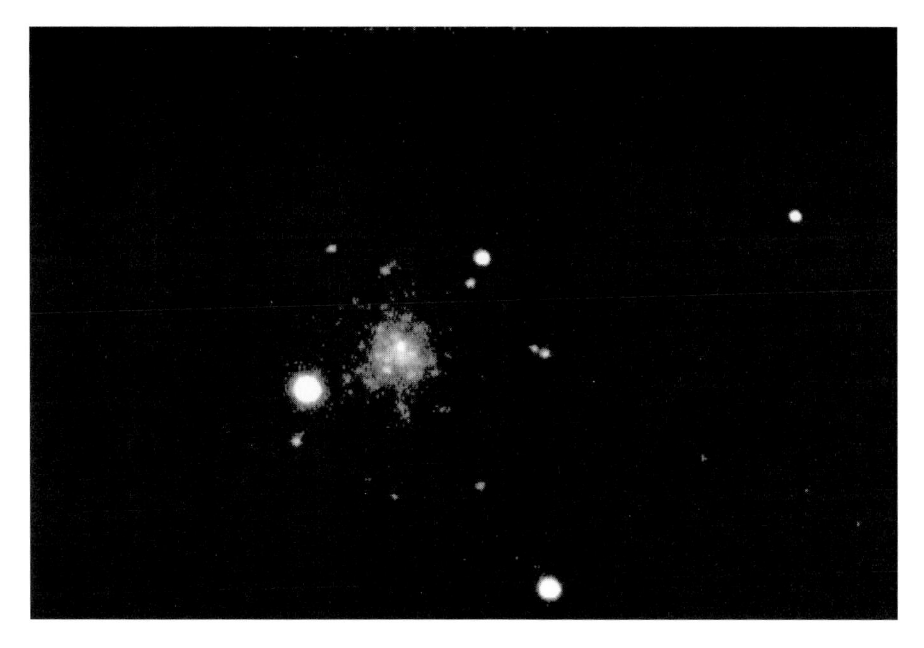

Figure 15.45. NGC 5634 © Tony O'Sullivan.

Figure 15.46. NGC 6229 © Tony O'Sullivan.

• NGC 5053

Other Name	Type	Const	Mag	Class
H 6.7	Globular	Coma Berenices	9.5	XI
RA	**Dec**	**Size**	**Distance**	**Rating**
13h 16m 27s	+17°41'53"	10'	53,500 Lyr	Hard

Description and Notes

With a low surface brightness, NGC 5053 is difficult to resolve, but the cluster is located only 1° South East from the bright globular M53 also in Coma, which should aid in detection. Observing with an 8" delivers a faint although quite large image without resolution but with larger optics, the component stars can be resolved. This globular is particularly loose and resembles a faint open cluster.

• NGC 5139

Other Name	Type	Const	Mag	Class
Omega Centauri	Globular	Centaurus	3.7	VIII
RA	**Dec**	**Size**	**Distance**	**Rating**
13h 26m 46s	–47° 28' 36"	55.0	18,250 Lyr	Easy

Figure 15.47. M12 © Tony O'Sullivan.

Description and Notes

No star cluster observing list would be complete without the largest, brightest and most spectacular globular, Omega Centauri. Unfortunately, it has a southerly declination, and cannot be seen from the UK, but is visible in southern parts of the USA. It is an easy object in binoculars, and may even be visible to the naked eye, and can be partially resolved with small 3 to 4" telescopes. Its member stars have an overall blue cast and are extremely dense, especially towards the core. NGC 5139 is very easy to find; look for a hazy patch between the two bright stars, zeta and gamma Centauri. A definite object to catch if travelling south on holiday.

• M3

Other Name	Type	Const	Mag	Class
NGC 5272	Globular	Canes Venatici	6.2	VI
RA	**Dec**	**Size**	**Distance**	**Rating**
13h 42m 11s	28°22'32"	18'	32,000 Lyr	Medium

Description and Notes

M3 is another fairly tricky cluster to find. In fact, from my suburban location I could not see it in my finder so used my "nudge and judge" method. Knowing M3 lay roughly halfway between Arcturus in Bootes, and Cor Caroli in Canes Venatici, I eventually found this nice bright globular. At 133× with my 8", I could not resolve this cluster fully but it did open up and show some mottled detail. From the UK, it is not really a naked eye object and is faint in binoculars, but does show a spherical haze with a 4" scope.

• NGC 5634

Other Name	Type	Const	Mag	Class
H 1.70	Globular	Virgo	9.5	IV
RA	**Dec**	**Size**	**Distance**	**Rating**
14h 29m 37s	–05°58'35"	5.5'	82,200 Lyr	Hard

Description and Notes

NGC 5634 is conveniently placed almost exactly between the stars mu and iota Virginis, so it should be easy to locate. However, at magnitude 9.5, this globular is not particularly bright, but as it is small, the surface brightness is condensed, which helps visibility. Not much resolution is detectable in an 8" scope, but increased aperture and magnification will show a bright mottled core and some faint outer halo stars.

Figure 15.48. M10 © Cliff Meredith.

• NGC 5897

Other Name	Type	Const	Mag	Class
H 6.8	Globular	Libra	8.5	XI
RA	**Dec**	**Size**	**Distance**	**Rating**
15h 17m 24s	–21° 00' 37"	11'	40,400 Lyr	Hard

Description and Notes

From northern latitudes, this globular is fairly low and can be affected by sky glow. In an 8" telescope the cluster is faint and diffuse, showing little detail and no resolution. Telescopes in the region of 10 to 12" should be able to resolve some of the member stars, though a larger telescope will show the full extent of this loose, large globular. NGC 5897 can be found roughly 5° North East of sigma Librae, lying almost in between this star and gamma Librae.

• M5

Other Name	Type	Const	Mag	Class
NGC 5904	Globular	Serpens	5.65	V
RA	**Dec**	**Size**	**Distance**	**Rating**
15h 18m 34s	+02° 04'58"	23'	25,000 Lyr	Easy

Description and Notes

A fairly large and bright globular, M5 is quite an easy target with a 4" telescope and shows some structure and detail, especially towards the outer edges. In a larger telescope, such as an 8" more resolution is evident and several chains of stars appear to radiate from its central regions. To find this cluster, locate alpha Serpentis and 109 Virginis, M5 lies roughly in between the two, more precisely, 7.5° away from alpha Serpentis. I personally cannot see any color within this globular, but reports of blues and yellows abound: what can you see?

• M80

Other Name	Type	Const	Mag	Class
NGC 6093	Globular	Scorpius	7.3	II
RA	**Dec**	**Size**	**Distance**	**Rating**
16h 17m 02s	–22°58'30"	10'	32,600 Lyr	Medium

Description and Notes

M80 is a rich, compressed, bright cluster, embedded in the Scorpion's "pincer," between the stars beta and sigma Scorpii, and has a small dense core. Although detectable in small telescopes, resolving the cluster is difficult, though averted vision may help. Alternatively, medium apertures 8" and above will resolve some of the magnitude 14–15 stellar components.

• M107

Other Name	Type	Const	Mag	Class
NGC 6171	Globular	Ophiuchus	7.9	X
RA	**Dec**	**Size**	**Distance**	**Rating**
16h 32m 32s	–13° 03'13"	13'	20,900 Lyr	Medium

Description and Notes

From the UK, M107 is quite low on the horizon and pales in comparison to nearby M10 and M12 also in Ophiuchus. At high magnification, this cluster appears reasonably loose and could be mistaken for a faint open cluster. Using my 4" refractor, I was able to detect but not resolve member stars, but larger telescopes and high magnification should achieve this and reveal several dark lanes. M107 is situated in between zeta and phi Ophiuchus.

• M13

Other Name	Type	Const	Mag	Class
NGC 6205	Globular	Hercules	5.8	V
RA	**Dec**	**Size**	**Distance**	**Rating**
16h 41m 42s	+36° 27'37"	20'	25,100 Lyr	Easy

Description and Notes

The brightest globular cluster in the Northern Hemisphere, M13 is an impressive spherical swarm. Easily located in the "keystone" asterism between eta and zeta Herculis this cluster in visible with the naked eye under very dark skies, but is easy in binoculars. In a 4" scope the object takes on a mottled structure, but an 8" is really required to resolve the cluster into a myriad of stars. A really large telescope can penetrate deep into the core stars and provided a breathtaking view. Large scopes and CCD images may also detect the famous "propeller" or "Y" shaped dust lane.

• NGC 6229

Other Name	Type	Const	Mag	Class
H 4.50	Globular	Hercules	9.4	IV
RA	**Dec**	**Size**	**Distance**	**Rating**
16h 46m 59s	+47°31'40"	4.5'	99,100 Lyr	Hard

Description and Notes

Of the three globular clusters in Hercules NGC 6229 is by far the faintest but still a worthwhile target. Situated approximately 4.5° North East of tau Herculis, this cluster is seen at its best in a large telescope, but an 8" can detect its faint

Figure 15.49. M92 © Cliff Meredith.

glow. However, do not expect to resolve this cluster as the member stars are magnitude 15–16.

• M12

Other Name	Type	Const	Mag	Class
NGC 6128	Globular	Ophiuchus	6.7	VII
RA	**Dec**	**Size**	**Distance**	**Rating**
16h 47m 14s	–01°56'52"	16'	16,000 Lyr	Easy

Description and Notes

M12 is situated approximately 3° North East of M10 and both should be visible at the same time, along with several other bright globulars in the vicinity, including M9. M12 itself is very bright and starts to resolve at 133×. It reminded me of M13 in Hercules in shape and brightness though obviously smaller. I also found M12 much brighter than nearby M10, even though they are of similar size and magnitude. To locate M12, move approximately 2° South from M10. Large telescopes will reveal much more stellar detail than visible in the 8".

• NGC 6231

Other Name	Type	Const	Mag	Class
Caldwell 76	Open	Scorpius	2.6	I 3 p n
RA	**Dec**	**Size**	**Distance**	**Rating**
16h 54m 10s	–41°49'30"	14'	4,052 Lyr	Easy

Description and Notes

NGC 6231 in Scorpius may be on the horizon limits for many Northern Hemisphere observers, but is an excellent binocular object if within your reach, showing about 10 stars between magnitude 5 and 6. The cluster is also attractive in small telescopes, and a 4" will reveal 15 or more members in a blue shroud of color reminiscent of the Pleiades. Located 0.5° North of zeta Scorpii, it is easily found.

• M10

Other Name	Type	Const	Mag	Class
NGC 6254	Globular	Ophiuchus	6.6	VII
RA	**Dec**	**Size**	**Distance**	**Rating**
16h 57m 9s	–04°05'58"	20'	14,400 Lyr	Easy

Description and Notes

At 48× in an 8" telescope, M10 is fairly bright and is easily located about 9° North East of zeta Ophiuchi, and 1° from magnitude 5 star 30 Ophiuchi. Increasing magnification to 96× the cluster started to resolve but only slightly. Going to an even higher power did not improve the view, probably due to seeing conditions and is

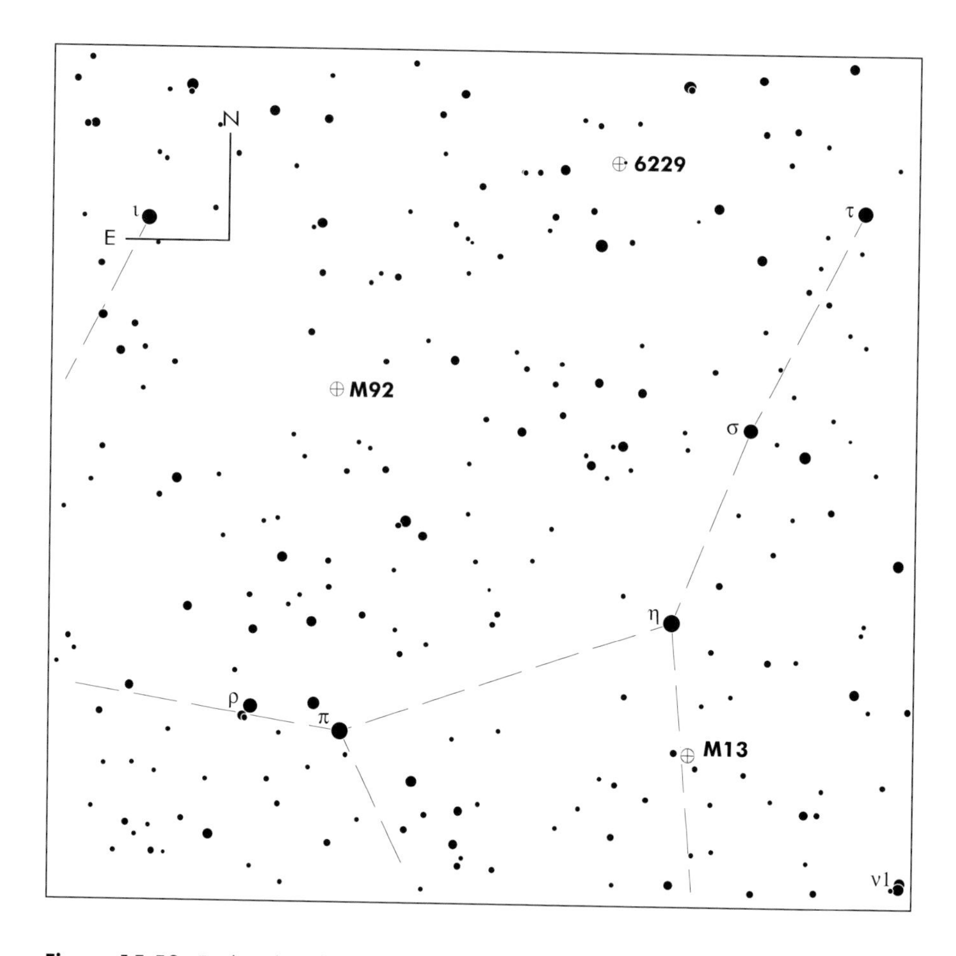

Figure 15.50. Finder Chart for Hercules © Software Bisque.

a case in point that higher magnifications are not always the most lucrative. Several chains of stars can be seen at these powers.

• M62

Other Name	Type	Const	Mag	Class
NGC 6266	Globular	Ophiuchus	6.5	IV
RA	**Dec**	**Size**	**Distance**	**Rating**
17h 01m 12s	–30°06'44"	15'	22,500 Lyr	Easy

Description and Notes

M62 is one of the brightest globulars in Ophiuchus. Unfortunately for Northern Hemisphere observers, it is also one of the lowest on the horizon. If this cluster is within your range, you will find it about 6° South East of Tau Scorpii. In small telescopes, M62 is visible but is faint and not resolved. Using an 8" telescope, a bright core is revealed and the asymmetrical shape of the cluster is presented, with some of the member stars resolved. Larger instruments show a pronounced mottling and several chains and strings of stars.

Figure 15.51. M9 © Tony O'Sullivan.

• M92

Other Name	Type	Const	Mag	Class
NGC 6341	Globular	Hercules	6.5	IV
RA	**Dec**	**Size**	**Distance**	**Rating**
17h 17m 07s	+43° 08'11"	14'	26,700 Lyr	Easy

Description and Notes

Although M92 is often overlooked in favor of the nearby M13, I found this cluster visually very appealing. It is bright and compact, very dense towards the core, and the outer stars fade off rapidly. The cluster resolved well in my 8" at a medium power of 133× and should be visible in binoculars as a spherical glow. A small telescope will reveal some structure, and a large telescope can really penetrate the stellar nucleus. M92 can be found about 7° North East of Eta Herculis, in the direction of iota Herculis.

• M9

Other Name	Type	Const	Mag	Class
NGC 6333	Globular	Ophiuchus	7.8	VIII
RA	**Dec**	**Size**	**Distance**	**Rating**
17h 19m 12s	–18°30'59"	12'	25,800 Lyr	Medium

Figure 15.52. M14 © Cliff Meredith.

Description and Notes

Even in an 8" telescope, at medium power M9 is quite faint and appears as a dim smudge of light. Using high power I could not resolve this cluster fully, and very little detail was apparent, although the cluster has an irregular, almost elliptical look. This cluster can be found about 3.5° South of the star eta Ophiuchi or Sabik and is relatively small and tightly compressed towards the core.

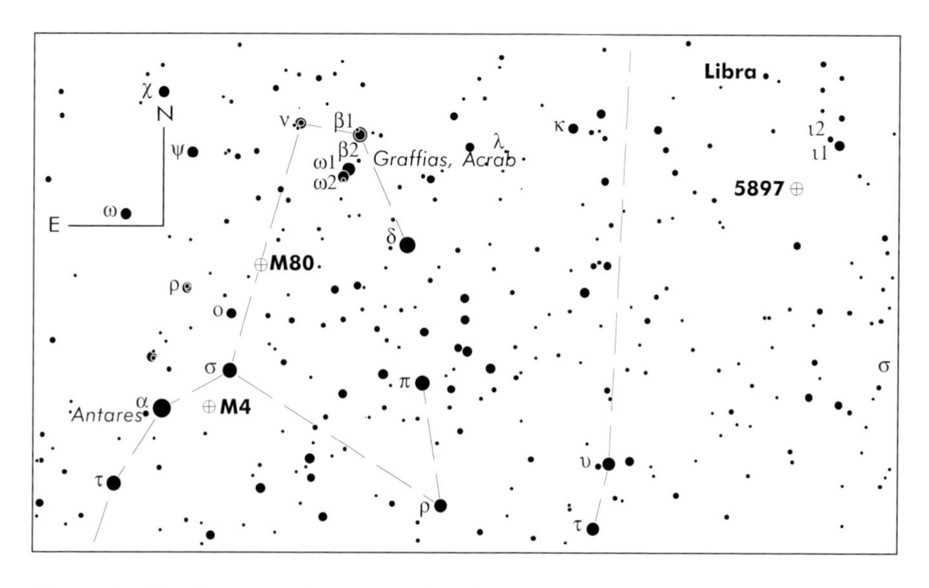

Figure 15.53. Finder Chart for Scorpius © Software Bisque.

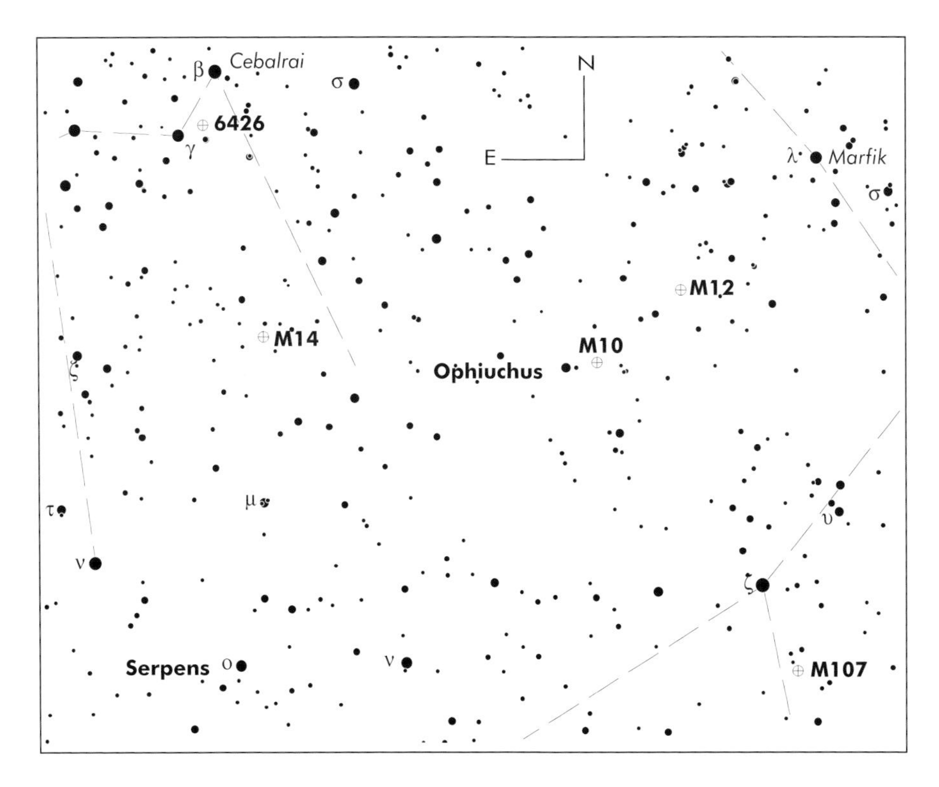

Figure 15.54. Finder Chart 1 for Ophiuchus © Software Bisque.

• NGC 6342

Other Name	Type	Const	Mag	Class
H 1.149	Globular	Ophiuchus	9.7	IV
RA	**Dec**	**Size**	**Distance**	**Rating**
17h 21m 10s	−19°35'14"	4.4	28,000 Lyr	Hard

Description and Notes

Ophiuchus contains over 20 globulars clusters of varying magnitudes and NGC 6342 is faint, but by no means the faintest. Small and medium optics reveal little in terms of detail or resolution but the cluster is detectable with an 8". A brighter core and more defined structure is visible in large scopes 16" and upward, displaying some faint stars and hints of dust lanes. To find this object point your scope 4.5° South East from eta Ophiuchi.

• NGC 6356

Other Name	Type	Const	Mag	Class
H 1.48	Globular	Ophiuchus	8.3	II
RA	**Dec**	**Size**	**Distance**	**Rating**
17h 23m 35s	−17°48'47"	7.2'	49,600 Lyr	Medium

Description and Notes

NGC 6356 is situated roughly 1° North East of M9, another globular in Ophiuchus, although M9 is brighter. Along with NGC 6342, all of three of these clusters can be seen in an 8" or larger telescope with a wide field of view. NGC 6356 is a fairly condensed object with a bright core and can be spotted in a smaller scope such as a 4" model. Large telescopes, for example a 16" display an elongated structure and some mottling but the cluster is not fully resolved.

• M14

Other Name	Type	Const	Mag	Class
NGC 6402	Globular	Ophiuchus	7.6	VIII
RA	**Dec**	**Size**	**Distance**	**Rating**
17h 37m 36s	–03° 14' 45"	11'	30,300 Lyr	Medium

Description and Notes

Another globular situated in Ophiuchus, M14 is similar to M9 in terms of it being faint and fairly small, but this object reveals a bit more detail at high power. Moving 7.5° South West of the star beta Ophiuchi should bring this cluster into view, but M14 resides in a sparse star field so there are no bright stars nearby. Binoculars do not give much away, but small scopes should improve the view. In an 8" this globular started to resolve at 133×, but at low power it could be mistaken for a comet or a galaxy.

• M6

Other Name	Type	Const	Mag	Class
NGC 6405	Open	Scorpius	4.2	III 2 p
RA	**Dec**	**Size**	**Distance**	**Rating**
17h 40m 20s	–32°15'12"	20'	15,87 Lyr	Easy

Description and Notes

Perhaps a little low for mid-Northern Hemisphere observers, however M6 is one of the brightest open clusters and covers a wide area of sky. Containing over a 100 stars, M6 can be seen with the naked eye under dark skies, and excels in small scopes at low power, where the clusters "butterfly" shape is evident. M6 is situated about 5.5° South West of gamma Saggitarii, and 4° South East of M7, another splendid magnitude 3.3 open cluster.

• NGC 6426

Other Name	Type	Const	Mag	Class
H 2.587	Globular	Ophiuchus	11.2	IX
RA	**Dec**	**Size**	**Distance**	**Rating**
17h 44m 54s	+03°10'13"	4.2'	67,500 Lyr	Difficult

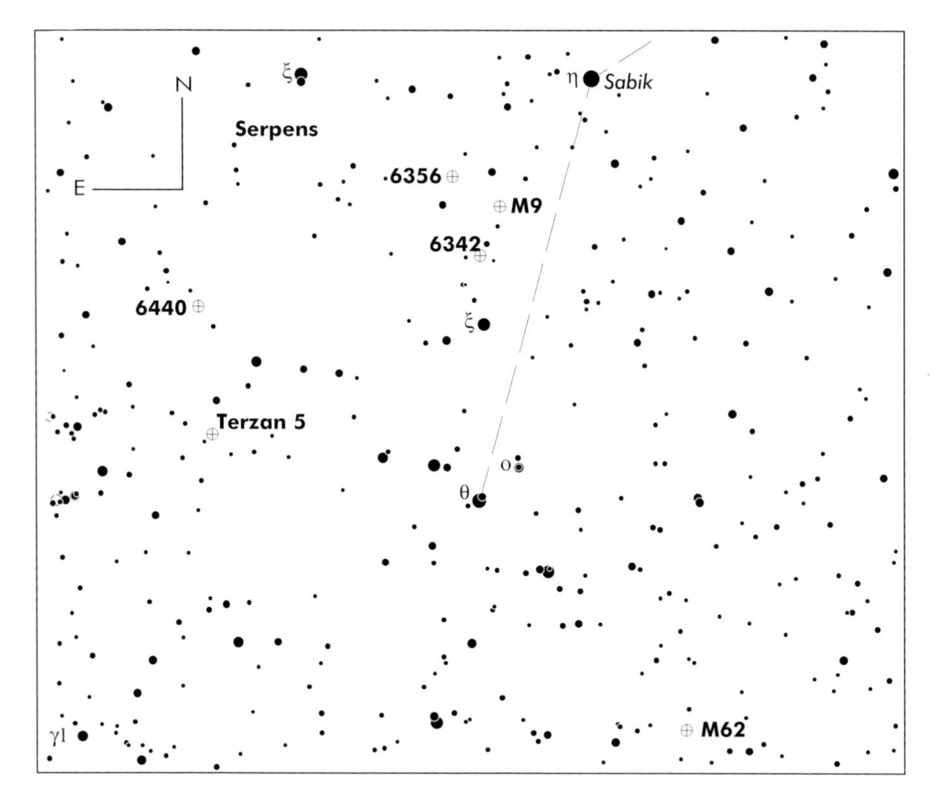

Figure 15.55. Finder Chart 2 for Ophiuchus © Software Bisque.

Description and Notes

Positioned less than 1° North West of gamma Ophiuchi, NGC 6426 would be easy to find except for the fact that it is very faint. Even with 8" optics only a small, spherical diffused object is visible and cannot be resolved. Moving up to a larger model, such as a 16" telescope improves the resolution and at high power the central core will be more evident and some mottling will be apparent. At this point, a few member stars can be resolved and the cluster should appear irregular in dimension.

• Terzan 5

Other Name	Type	Const	Mag	Class
Ter 5	Globular	Sagittarius	13.5	–
RA	**Dec**	**Size**	**Distance**	**Rating**
17h 48m 25s	–24°47'14"	2.1'	33,600 Lyr	Difficult

Description and Notes

This faint and tiny globular is located in a populated region of Sagittarius that contains many bright and faint, open and globular clusters. Terzan 5 is the brightest

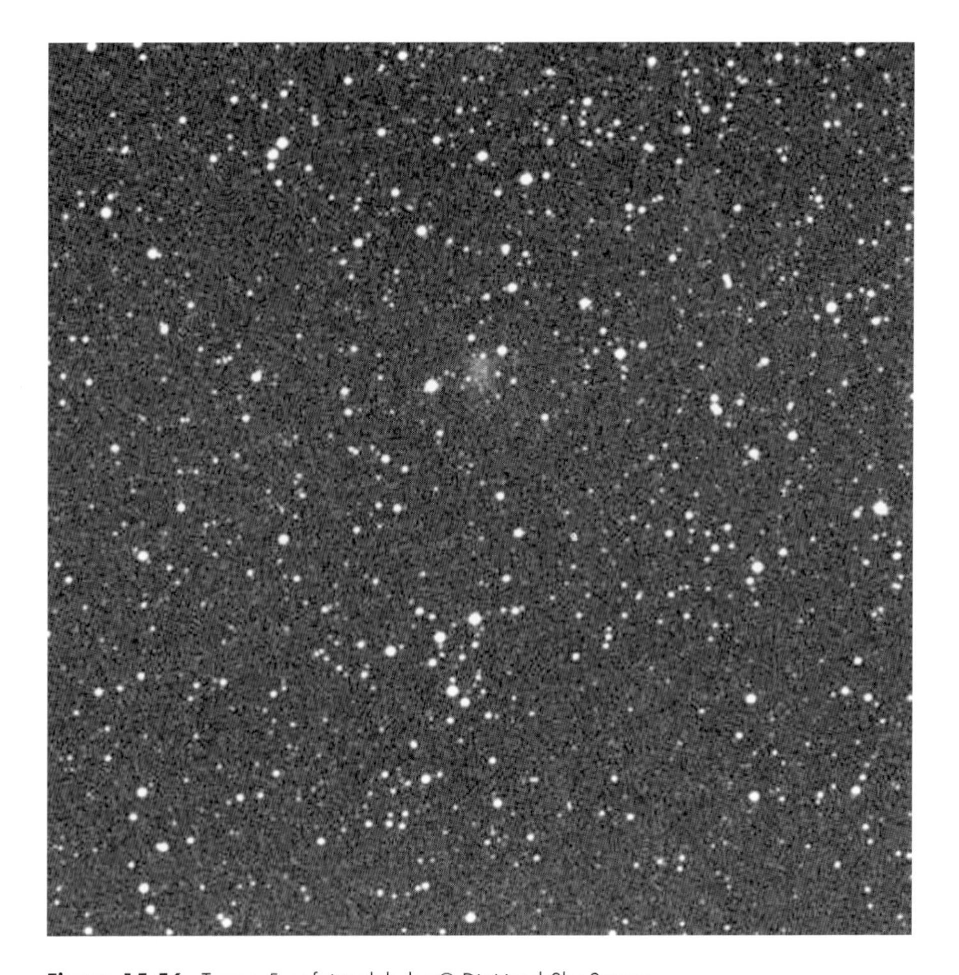

Figure 15.56. Terzan 5, a faint globular © Digitized Sky Survey.

member of the 11 strong Terzan catalogue, but this object is difficult to detect with anything less than a 10" telescope, and at this aperture little detail can be observed. With increased aperture of perhaps 12 to 16" the cluster contains some structure. Terzan 5 is situated just over 3° North of the star 3 Sagittarii.

• NGC 6440

Other Name	Type	Const	Mag	Class
H 1.150	Globular	Sagittarius	9.2	V
RA	**Dec**	**Size**	**Distance**	**Rating**
17h 48m 53s	–20°21'34"	4.4	27,400 Lyr	Hard

Description and Notes

As the component stars of this globular are a mere magnitude 17, resolving this cluster is difficult, and in an 8" telescope you will only detect a brighter core with

Figure 15.57. M16 Open Cluster and Nebula © Cliff Meredith.

Figure 15.58. M17 Open Cluster and Nebula © Cliff Meredith.

Figure 15.59. M26 © Tony O'Sullivan.

a concentrated nucleus. Situated roughly 7.5° North East of theta Ophiuchi, a 16" telescope will reveal a diffused halo along with some granular structure.

• NGC 6535

Other Name	Type	Const	Mag	Class
–	Globular	Serpens	10.5	XI
RA	**Dec**	**Size**	**Distance**	**Rating**
18h 03m 51s	–00°17'49"	3.4'	22,200 Lyr	Hard

Description and Notes

In Serpens, approximately 3° South from 67 Ophiuchi, this small, rather faint globular is located but will be difficult in scopes under 8". This size telescope can theoretically detect this object but under mediocre skies you may struggle. However, a large instrument will render some of the member stars and enable detail to be resolved, such as the irregular shape and grainy presentation.

• M16

Other Name	Type	Const	Mag	Class
NGC 6611	Open	Serpens	6.0	II 3 m n
RA	**Dec**	**Size**	**Distance**	**Rating**
18h 18m 48s	–13°48'24"	6'	5,700 Lyr	Easy

Description and Notes

M16 is actually a nebula with an embedded open cluster, and readers may be more familiar with its other designation, the Eagle Nebula, which is an amazing yet diffuse emission nebula. The open cluster itself is much easier to detect, contains about 60 stars and can be spotted in 4" telescopes, 2.5° North West of gamma Scuti, a magnitude 4 star.

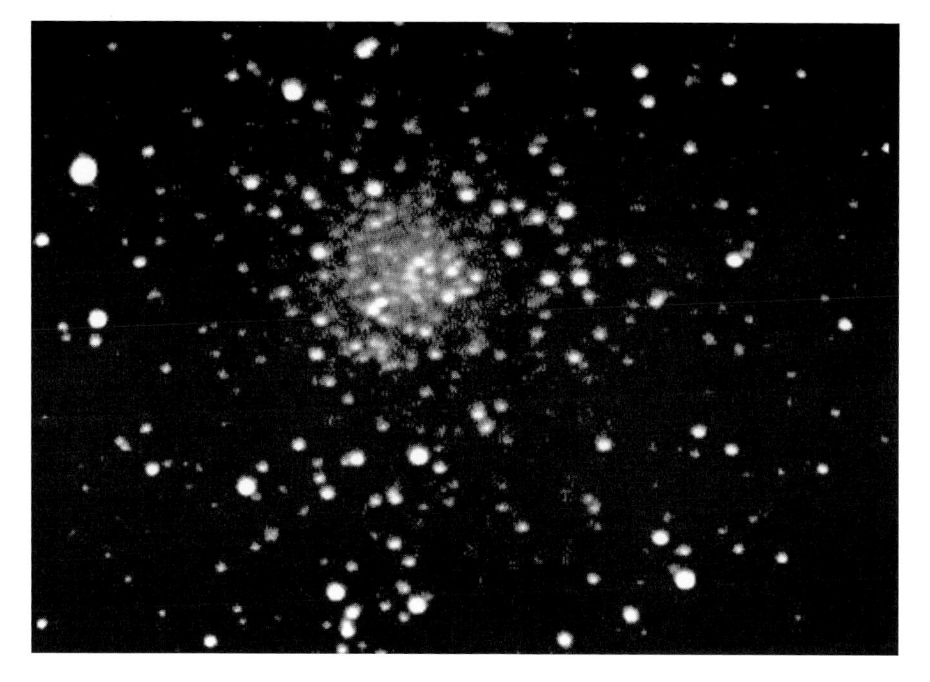

Figure 15.60. NGC 6712 © Tony O'Sullivan.

• M17

Other Name	Type	Const	Mag	Class
NGC 6618	Open	Sagittarius	6.0	III 3 m n
RA	**Dec**	**Size**	**Distance**	**Rating**
18h 20m 47s	−16°10'18"	25'	4,238 Lyr	Easy

Description and Notes

In a similar fashion to M16, this cluster is also associated with an emission nebula, with the common name of the "Swan Nebula." Using binoculars you may detect some of the 40 stars, but the brightest member is magnitude 9, so a small or medium telescope will stand you more chance. M16 is situated 2.5° South West of gamma Scuti.

• M22

Other Name	Type	Const	Mag	Class
NGC 6656	Globular	Sagittarius	5.1	VII
RA	**Dec**	**Size**	**Distance**	**Rating**
18h 36m 24s	−23°54 ' 12"	32'	10,400 Lyr	Easy

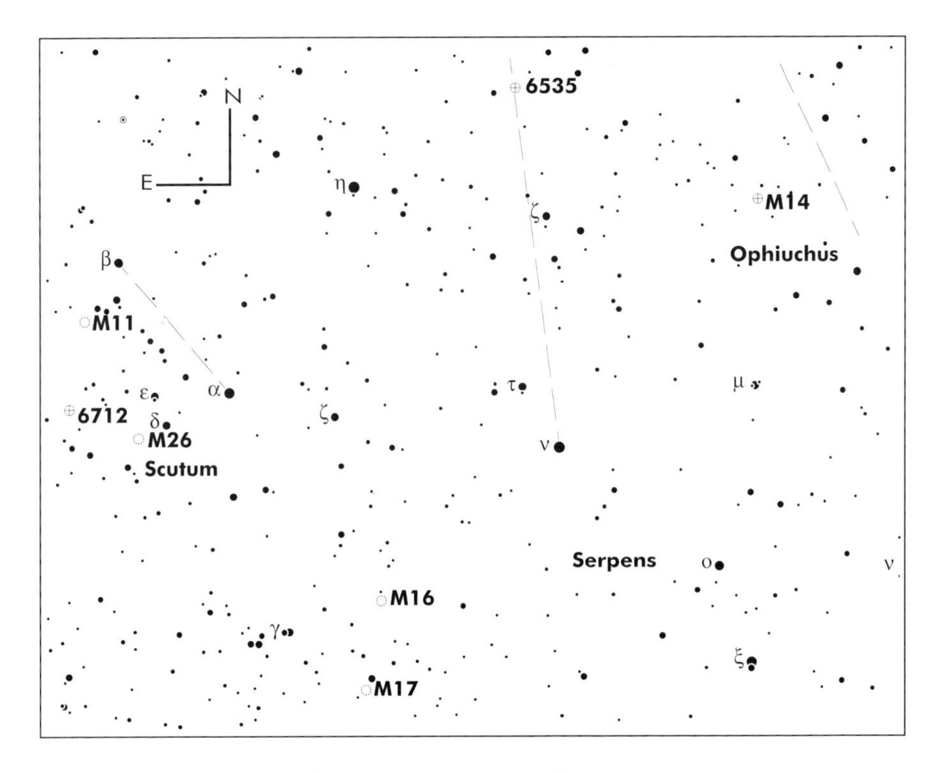

Figure 15.61. Finder Chart for Scutum & Serpens © Software Bisque.

Description and Notes

M22 is another fine example of a large bright globular cluster that can be seen with most instruments and detected with binoculars. It is technically a Southern Hemisphere object but can be observed from some mid-northern latitudes, albeit low in the skyline. As the brightest globular in Sagittarius it is easy to spot, about 2° North East from the "teapot lid" star lambda Borealis. With member stars from magnitude 11, this cluster is the third brightest globular in the sky, one of the closest to us, and resolves partially even in a 4" scope as a slightly oval-shaped globe. Larger telescopes will reach further into its stellar core where you may see several dust lanes.

• M26

Other Name	Type	Const	Mag	Class
NGC 6694	Open	Scutum	8.0	I 1 m
RA	**Dec**	**Size**	**Distance**	**Rating**
18h 45m 18s	–09°23 ' 00"	7'	5,216 Lyr	Medium

Description and Notes

M26 is fairly close to M11, the Wild Duck Cluster, but this object is much fainter and smaller than M11, looking like a small globular at low power. More magnifi-

Figure 15.62. Palomar 9 © Digitized Sky Survey.

cation shows an almost asterism like cluster in my 4" refractor, situated about 2.5° South East of alpha Scuti. In the same telescope I managed to detect about 18 members though a larger aperture would improve this count.

• M11

Other Name	Type	Const	Mag	Class
NGC 6705	Open	Scutum	5.8	I 2 r
RA	**Dec**	**Size**	**Distance**	**Rating**
18h 51m 05s	–06°16 ' 12"	13'	6,119 Lyr	Easy

Description and Notes

M11, also known as the "Wild Duck Cluster" is bright and resolved easily at 48× with a 4" telescope. The cluster actually contains over 500 member stars up to magnitude 14, but most are brighter than this figure. M11 is quite a dense object for an open cluster and resembles a globular at low power, and with an 8" instrument I detected over 50 stars. To find this cluster, move just under 2° South East of beta Scuti.

• NGC 6712

Other Name	Type	Const	Mag	Class
H 1.47	Globular	Scutum	8.1	IX
RA	**Dec**	**Size**	**Distance**	**Rating**
18h 53m 04s	–08 °42 ' 22"	9.8'	22,500 Lyr	Medium

Description and Notes

NGC 6712 is difficult to resolve in an 8" telescope, or even at high power, although some grainy structure can be seen. Larger telescopes 10 to 12" can resolve about two dozen stars and the cluster takes on a more irregular form. Very large telescopes 16" plus, however can reveal much more detail especially in the core region. NGC 6712 is positioned approximately 5° South West of lambda Aquilae.

• Palomar 9

Other Name	Type	Const	Mag	Class
NGC 6717	Globular	Sagittarius	9.3	VIII
RA	**Dec**	**Size**	**Distance**	**Rating**
18h 55m 6s	–22°42 ' 03"	5.4'	23,100 Lyr	Difficult

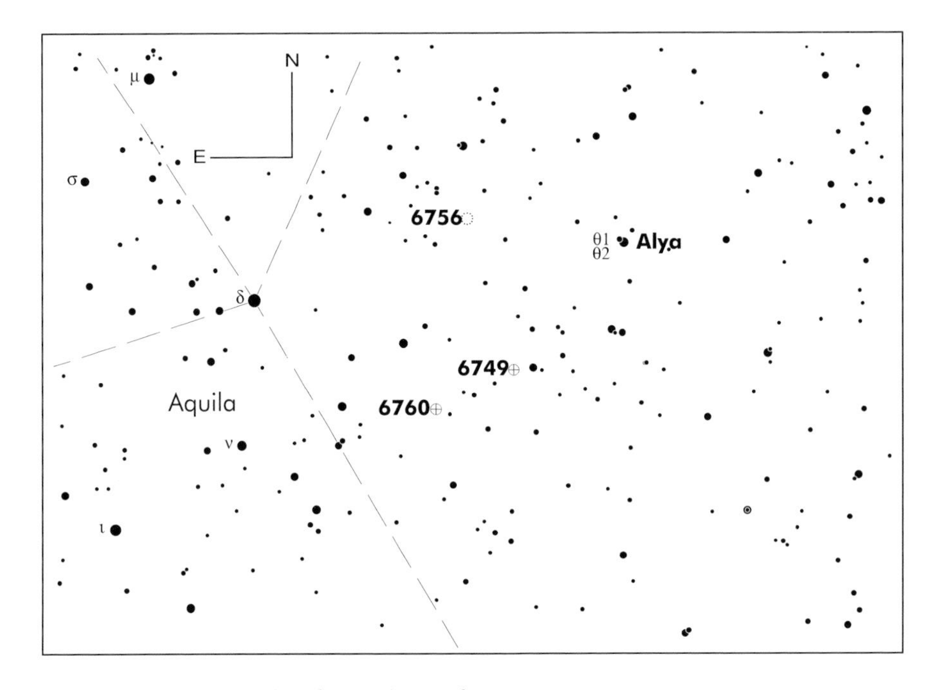

Figure 15.63. Finder Chart for Aquila © Software Bisque.

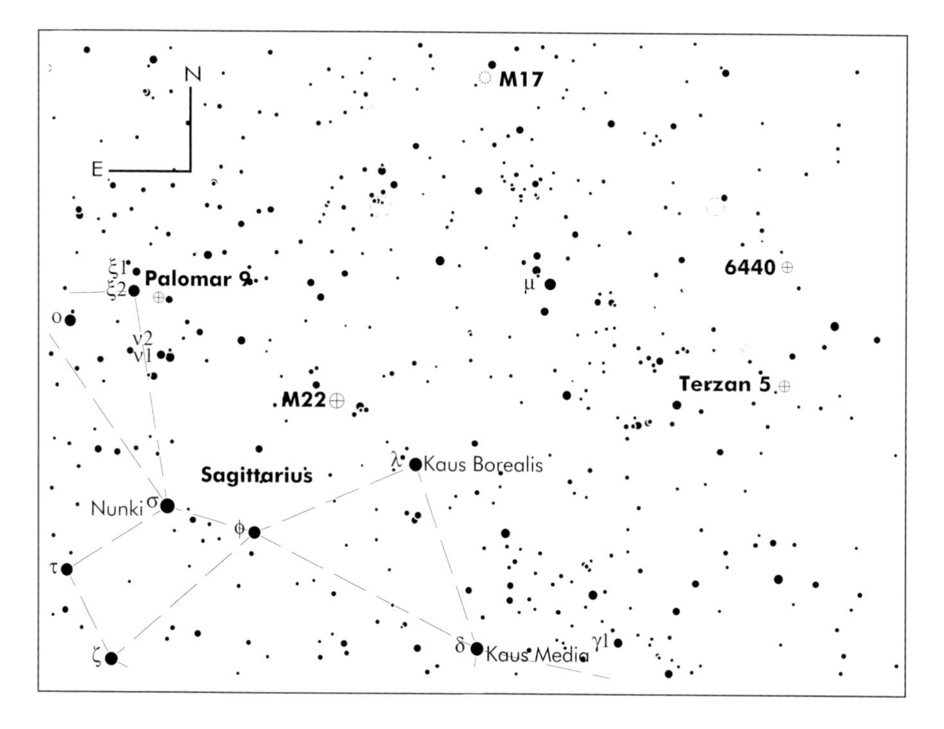

Figure 15.64. Finder Chart for Sagittarius © Software Bisque.

Description and Notes

Palomar 9 is the brightest member of the Palomar group of faint globulars, and is located approximately 3° West of pi Sagittarii. With a 12" telescope, this cluster appears small and diffused with some member stars resolved. A 16" scope reveals a small bright nucleus and further stars, though some of these are superimposed on the cluster itself. It is not an object for small or medium scopes unless a CCD or similar unit is used.

• NGC 6749

Other Name	**Type**	**Const**	**Mag**	**Class**
Berkeley 42	Globular	Aquila	12.45	IX
RA	**Dec**	**Size**	**Distance**	**Rating**
19h 05m 15s	+01°54' 03"	4'	25,800 Lyr	Difficult

Description and Notes

NGC 6749 a much fainter globular than nearby NGC 6760, has a very low surface brightness and is extremely small. As the cluster lies in a dense region of the Milky Way, its visibility is further reduced. In a 12" telescope an extremely dim smudge can be detected with a few non member stars seen with averted vision. A much larger class of telescope, 16" plus is required to show modest brightening of the

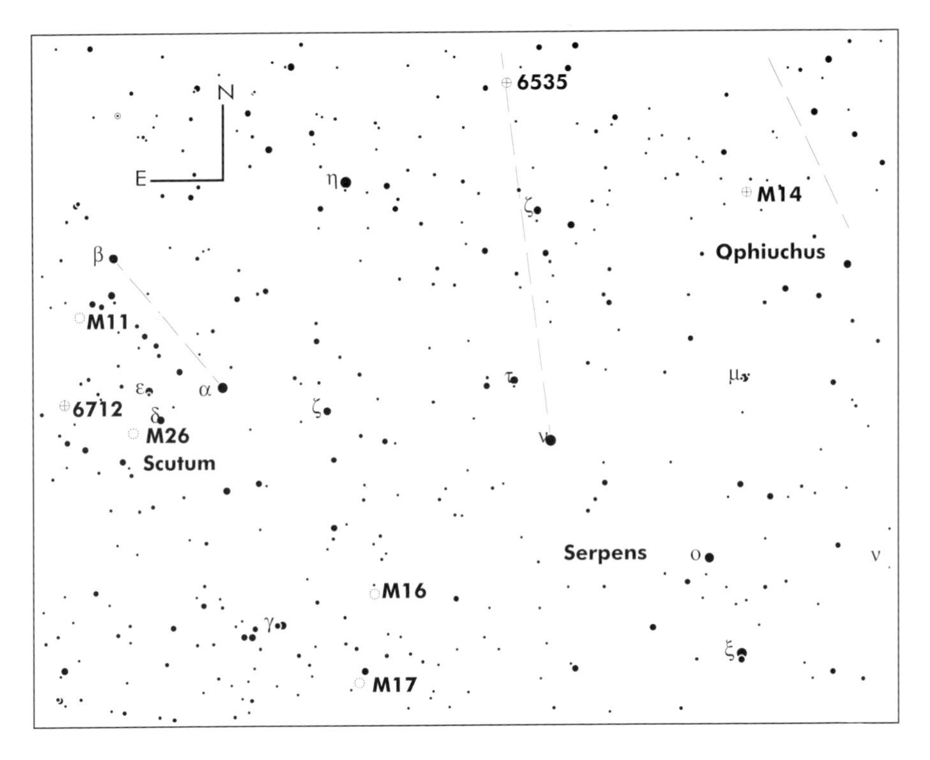

Figure 15.65. Finder Chart for Serpens & Scutum © Software Bisque.

Figure 15.66. NGC 6760 © Tony O'Sullivan.

core and a slight irregular shape. NGC 6749 is one of the faintest globulars in the entire NGC catalogue so is only suitable for big scopes or CCD imaging.

• NGC 6756

Other Name	Type	Const	Mag	Class
OCL 99	Open	Aquila	10.6	I 2 m
RA	**Dec**	**Size**	**Distance**	**Rating**
19h 08m 42s	+04°42 ' 18"	4'	4,912 Lyr	Hard

Description and Notes

NGC 6756 is a tiny cluster only 4' wide and contains about 40 stars. Although the brightest member is a lowly magnitude 13; therefore a medium or large telescope, 8" plus is required for a good view. 16" of aperture reveals about two dozen members and several tightly packed knots of stars. The cluster is located 4.5° North West of delta Aquilae.

• NGC 6760

Other Name	Type	Const	Mag	Class
–	Globular	Aquila	8.9	IX
RA	**Dec**	**Size**	**Distance**	**Rating**
19h 11m 21s	+01°01' 50"	9.6'	24,100 Lyr	Medium

Description and Notes

There are only two globular clusters in Aquila (three if you include GLIMPSE-C01, a recent but very faint discovery) and the brighter cluster NGC 6760 is situated 4° South West of delta Aquila. Visually, this object is small and diffused in an 8" telescope and cannot be resolved. Increasing the aperture will improve the resolution, also enabling the possibility to detect NGC 6749 only 1.5° to the North West, but at magnitude 12.5 this is a very faint cluster.

• M56

Other Name	Type	Const	Mag	Class
NGC 6779	Globular	Lyra	8.3	V
RA	**Dec**	**Size**	**Distance**	**Rating**
19h 16m 35s	+30° 11' 05"	8.8	32,900 Lyr	Medium

Description and Notes

An easily found globular cluster, almost centered between beta Cygni (Albireo) and gamma Lyrae, M56 appears faint at low power. This is another globular that resembles a comet, but higher magnification (133×) begins to resolve the cluster with an 8" scope. Its features are delicate and subtle, and it is not as highly

Figure 15.67. M56 © Tony O'Sullivan.

concentrated towards the core as other globulars. Large telescopes will render more detail and uncover some dark lanes within the cluster.

• Collinder 399

Other Name	Type	Const	Mag	Class
Brocchi's cluster	Asterism	Vulpecula	3.6	–
RA	**Dec**	**Size**	**Distance**	**Rating**
19h 25m 4s	+20°11' 00"	90'	420 Lyr	Easy

Description and Notes

This infamous stellar group is also known as the "Coathanger" and when observed through binoculars or a small telescope you will see why – it certainly lives up to its name. Once thought to be an open cluster, this asterism is a collection of ten or so stars between magnitude 5 and 7 and is located 4° North West of alpha Sagittae.

• M71

Other Name	Type	Const	Mag	Class
NGC 6838	Globular	Sagitta	8.2	XI
RA	**Dec**	**Size**	**Distance**	**Rating**
19h 53m 46s	+18°46' 42"	7.2'	13,000 Lyr	Medium

Figure 15.68. The Coathanger, Collinder 399 © Tony O'Sullivan.

Figure 15.69. M71 © Tony O'Sullivan.

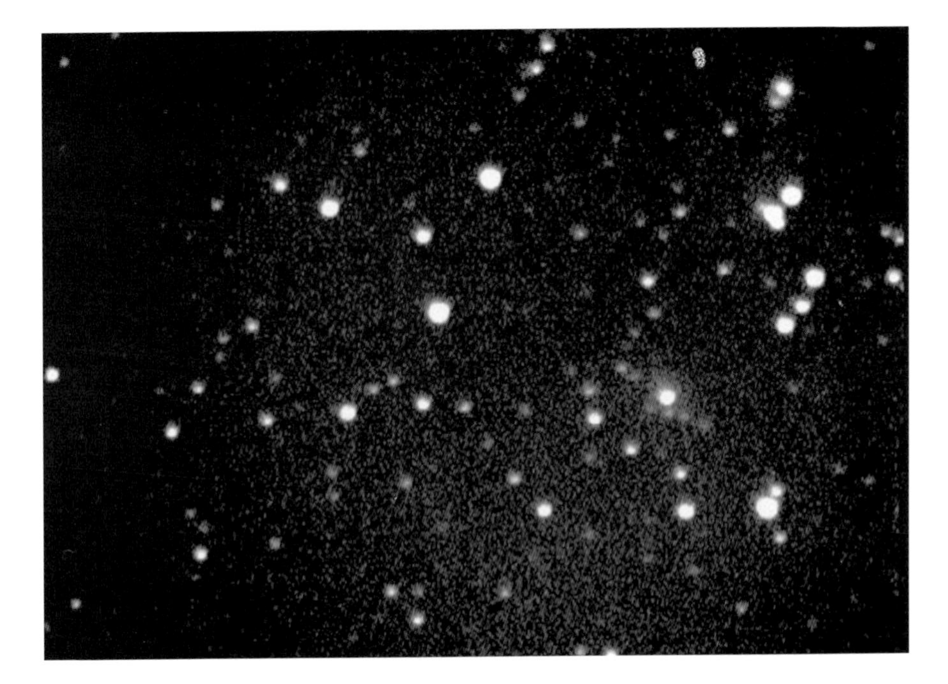

Figure 15.70. Roslund 4 and Faint Nebula IC 4954 © Tony O'Sullivan.

Description and Notes

This fairly faint globular appears nebulous at 48× in my 8" telescope, though does begin to resolve at 133×. M71 can easily be found between gamma and delta Saggitae and is quite a loose collection of stars with no dense central region. The classic "arrowhead" shape is more apparent in larger telescopes as the member stars are magnitude 11 to 16. From suburban skies or with mediocre weather, it does not resolve well.

• Roslund 4

Other Name	Type	Const	Mag	Class
–	Open	Vulpecula	10	IV 2 p n
RA	**Dec**	**Size**	**Distance**	**Rating**
20h 04m 54s	+29°13' 00"	5'	6,520 Lyr	Hard

Description and Notes

At magnitude 10, Roslund 4 is a pretty faint cluster situated 7.5° East of beta Cygni (Albireo). The cluster contains about 30 genuine member stars; however, the most luminous member is magnitude 11.6, so medium or large telescopes, 8- to 10" plus will provide the best views. Roslund 4 is part of the nebula IC 4954, which is very faint but can be picked up with a CCD.

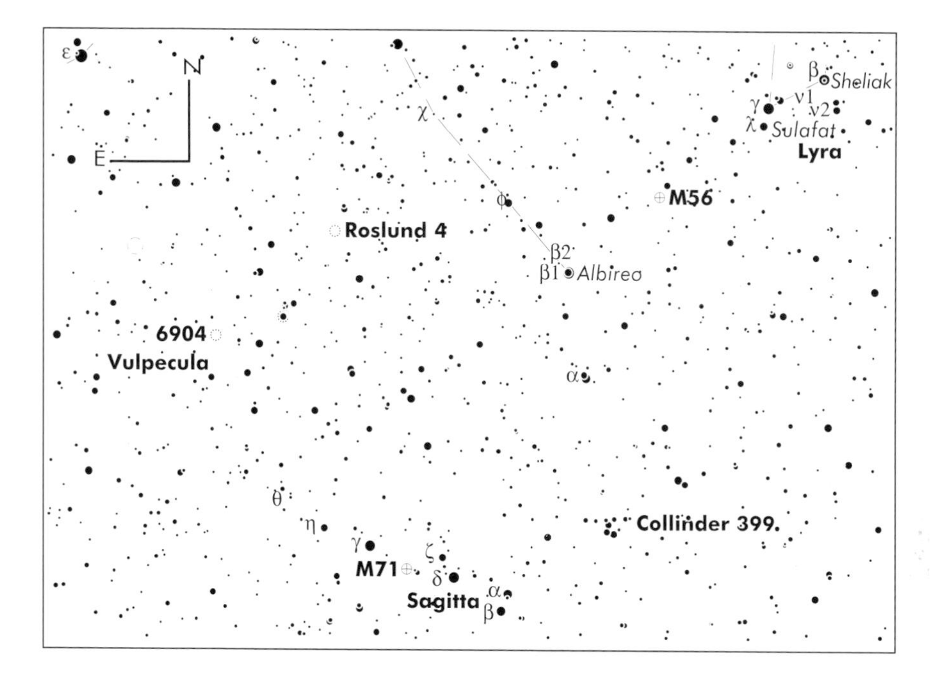

Figure 15.71. Finder Chart for Vulpecula, Lyra & Sagitta © Software Bisque.

Figure 15.72. IC 4996 © Tony O'Sullivan.

• IC 4996

Other Name	Type	Const	Mag	Class
–	Open	Cygnus	7.3	I 3 p n
RA	**Dec**	**Size**	**Distance**	**Rating**
20h 16m 30s	+37°38' 00"	6'	5,646 Lyr	Medium

Description and Notes

Although 15 stars from magnitude 8.5 are listed for this cluster, 3° South of gamma Cygni, only about 5 stars are easily seen towards its core, with the rest of the field containing mainly faint stars. A 4" can detect this object in good conditions, but a 6 to 8" will do a better job of resolving its stellar contents.

• NGC 6904

Other Name	Type	Const	Mag	Class
–	Open	Vulpecula	15.5	–
RA	**Dec**	**Size**	**Distance**	**Rating**
20h 21m 48s	+25°44' 24"	8'	–	Difficult

Figure 15.73. NGC 6904 © Tony O'Sullivan.

Figure 15.74. M29 © Tony O'Sullivan.

Description and Notes

NGC 6904 is described as non-existent in the RNGC, and listed as an asterism in several catalogues, but is in fact an open cluster, as shown in the latest open cluster catalogue (Dias). It contains about 24 stars from magnitude 9 to 13 and requires a very large telescope or CCD equipment to observe. It is also difficult to track down as there are no bright stars nearby, but is situated 1.5° East of the magnitude 5 star SAO 88410.

• M29

Other Name	Type	Const	Mag	Class
NGC 6913	Open	Cygnus	6.6	III 3 p n
RA	**Dec**	**Size**	**Distance**	**Rating**
20h 23m 57s	+38°30' 30"	10'	3,742 Lyr	Easy

Description and Notes

Approximately 1.5° South of gamma Cygni, the open cluster M29 can be detected in a small telescope and contains approximately 50 stars with the brightest member at magnitude 8.6. With my 8" instrument I resolved 8 or 9 members at 133×, with the main stars forming a trapezoid. A larger telescope of 12" can resolve over 20 stars in this loose cluster.

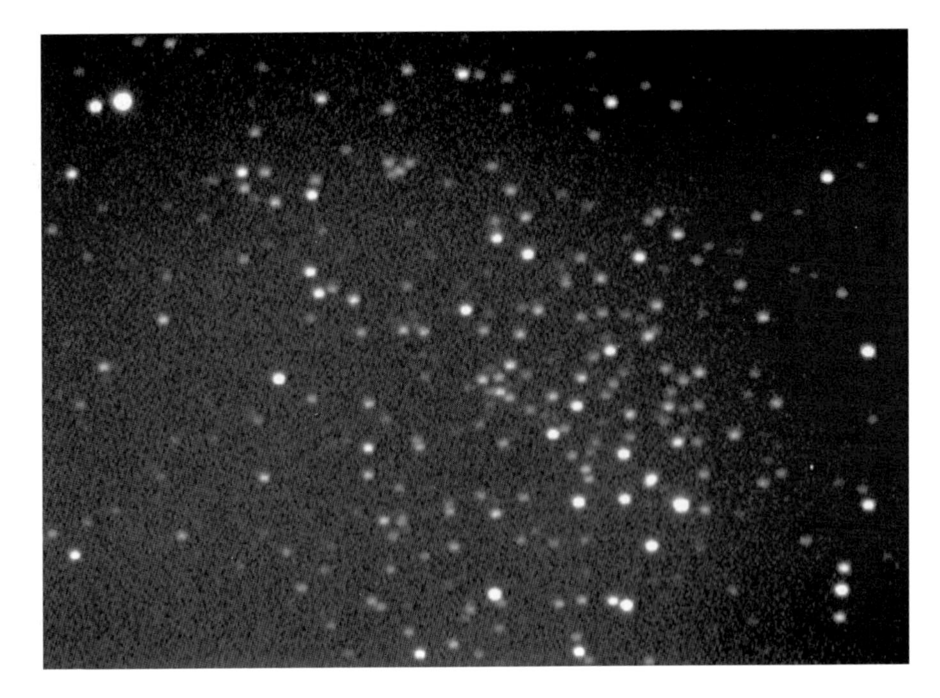

Figure 15.75. NGC6939 © Tony O'Sullivan.

• NGC 6939

Other Name	Type	Const	Mag	Class
OCL 217	Open	Cepheus	7.8	I 1 m
RA	**Dec**	**Size**	**Distance**	**Rating**
20h 31m 30s	+60°39' 42"	10'	5,868 Lyr	Hard

Description and Notes

NGC 6939 is a small, very faint cluster, located about 2° South West of eta Cephei. In my 4" refractor I could only just see a few component stars that were magnitude 9 and 10, contained by 4 brighter stars resembling a kite shape. This cluster would really benefit from a larger aperture where more of the 80 stellar members can be resolved.

• NGC 6934

Other Name	Type	Const	Mag	Class
H 1.103	Globular	Delphinus	8.9	VIII
RA	**Dec**	**Size**	**Distance**	**Rating**
20h 34m 12s	+07° 24' 15"	7.1	51,200 Lyr	Medium

Figure 15.76. NGC 6934 © Tony O'Sullivan.

Description and Notes

I found this cluster just over 10° East of alpha Aquilae (Altair) and it is close to another globular NGC 7006, though I found this one to be more luminous. It appeared somewhat bright and circular in the 8" at 133× but I could not fully resolve the cluster, although some mottled structure was evident at 266×.

• M72

Other Name	Type	Const	Mag	Class
NGC 6981	Globular	Aquarius	9.2	IX
RA	**Dec**	**Size**	**Distance**	**Rating**
20h 53m 30s	−12°32' 13"	6.6'	55,400 Lyr	Hard

Description and Notes

Just over 3° South East from epsilon Aquarii we find the globular M72, which is a pretty loose cluster with its member stars shining at magnitude 14. This cluster is actually the faintest Messier globular and is difficult to resolve in small telescopes. The core is dense at low power but structure is visible with a larger telescope and high magnification.

Figure 15.77. NGC 7006 © Tony O'Sullivan.

• NGC 7006

Other Name	Type	Const	Mag	Class
H 1.52	Globular	Delphinus	10.6	I
RA	**Dec**	**Size**	**Distance**	**Rating**
21h 01m 30s	+16°11' 15"	3.6	135,400 Lyr	Hard

Description and Notes

Observing this fairly bright globular presents a small circular haze with some brightening towards the clusters center, though I could not resolve its component stars with an 8," even at high power. Small and medium telescopes should detect this cluster, but do not expect to see any detail, even large telescopes struggle to improve the view. To track down this distant globular, head just over 3° East of gamma Delphini.

• M15

Other Name	Type	Const	Mag	Class
NGC 7078	Globular	Pegasus	6.2	IV
RA	**Dec**	**Size**	**Distance**	**Rating**
21h 29m 58s	+12 °10' 01"	18'	33,600 Lyr	Easy

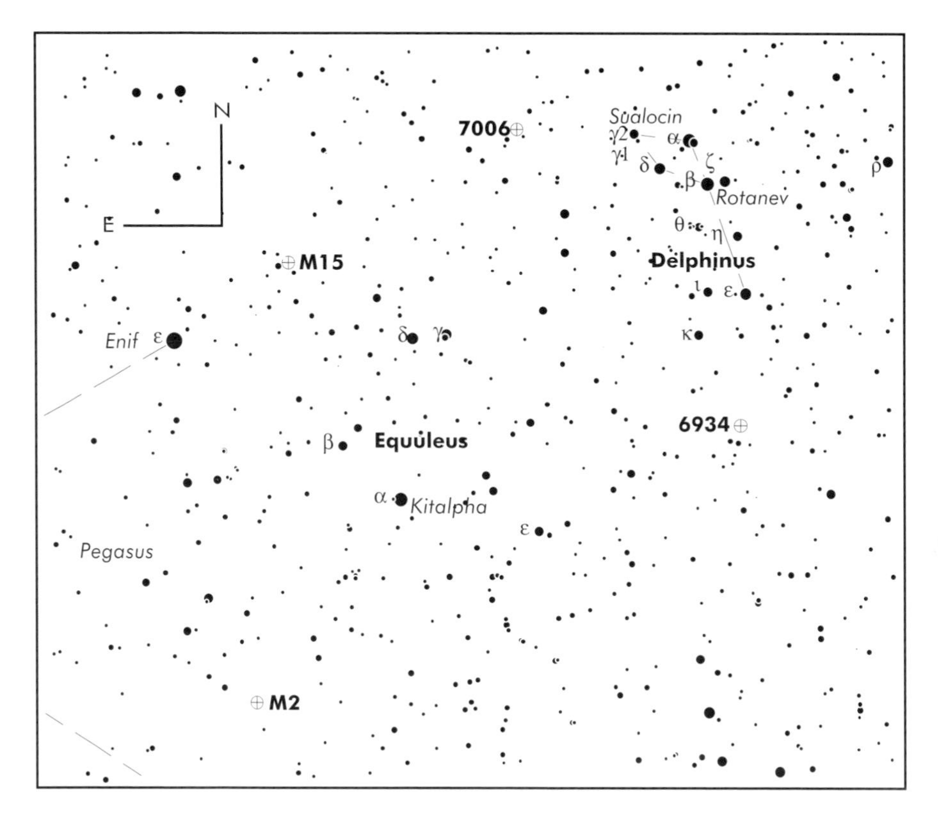

Figure 15.78. Finder Chart for Delphinus Pegasus & Equuleus© Software Bisque.

Figure 15.79. M15, a bright globular © Tony O'Sullivan.

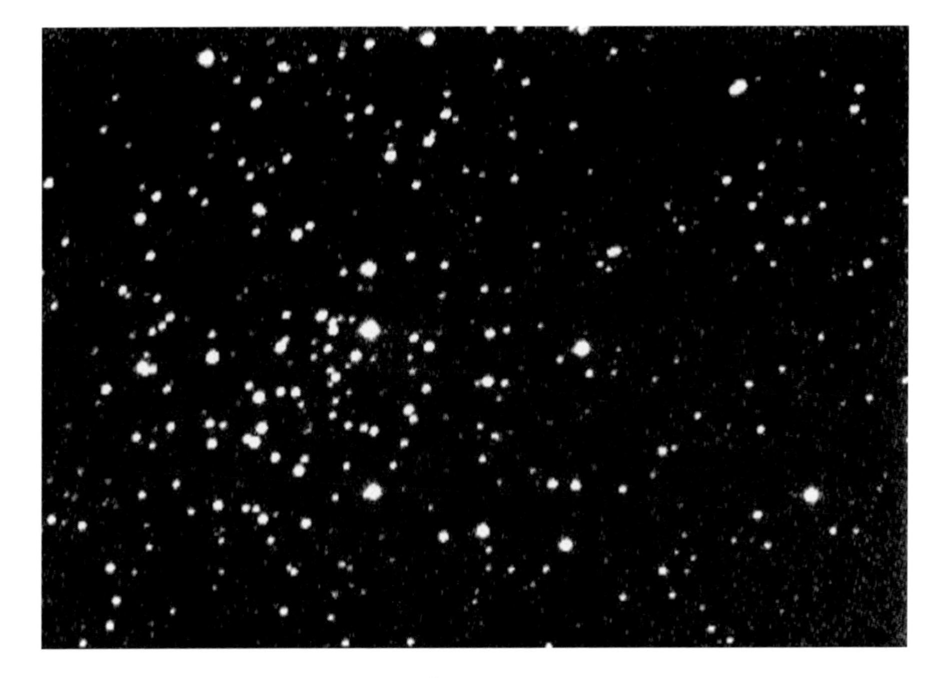

Figure 15.80. NGC 7086 © Tony O'Sullivan.

Description and Notes

M15 is one of my favorite globular clusters, and I prefer it in many ways to M13, which is actually larger and brighter! This object is more compact, though fainter than M13 and looks slightly irregular in shape. It can be detected with binoculars and small scopes add more detail to the view. However, in medium telescopes 8" and larger the cluster can be nicely resolved and shows a bright central region. Tracking down M15 is quite straightforward but requires a little star-hopping. Start at alpha Pegasi, one of the stars in the "Square" of Pegasus, then move down the "leg" to zeta, then onto theta Pegasi. At this point, head about 7° North West to epsilon Pegasi, and keep going in the same direction for another 4° and you should hit M15. Believe me, it is worth it!

• NGC 7086

Other Name	Type	Const	Mag	Class
OCL 214	Open	Cygnus	8.4	II 2 m
RA	**Dec**	**Size**	**Distance**	**Rating**
21h 30m 27s	+51°36' 00"	12'	4,231 Lyr	Hard

Description and Notes

There are approximately 50 stars associated with this open cluster, but the brightest member is only magnitude 10.2, so the figure for the actual cluster can be misleading.

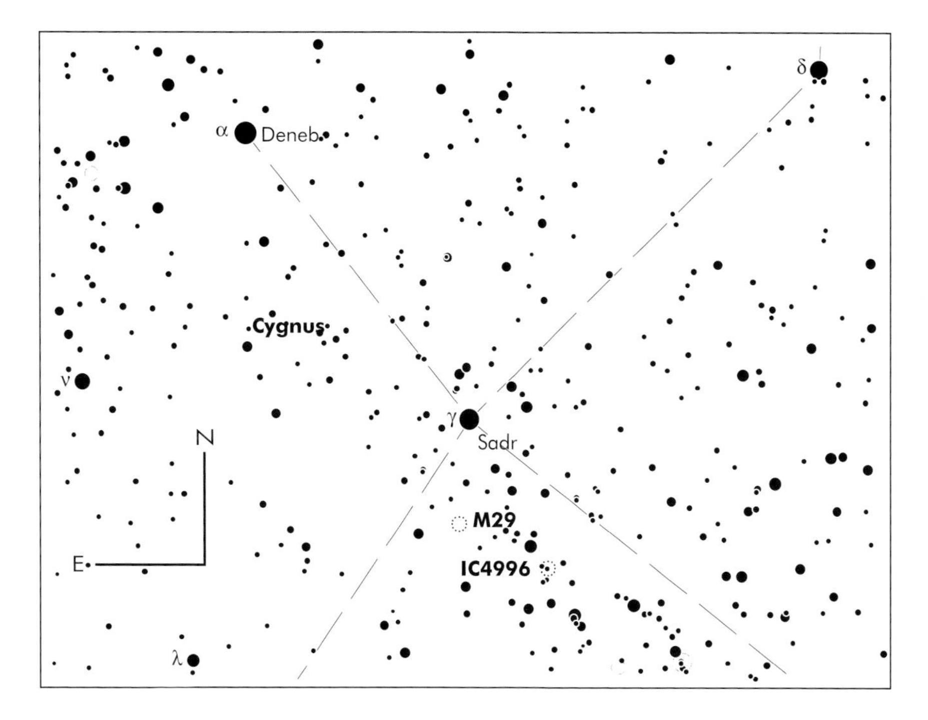

Figure 15.81. Finder Chart for Cygnus © Software Bisque.

Situated just under 2° West of magnitude 4 star pi Cygni, NGC 7086 is detectable in a small telescope, but an 8" instrument will show about 20 members. A larger 16" can discern about 70 stars in a rich compact group, including several double stars and prominent dark lanes. This cluster also makes a fine CCD target.

• M39

Other Name	Type	Const	Mag	Class
NGC 7092	Open	Cygnus	4.6	III 2 p
RA	**Dec**	**Size**	**Distance**	**Rating**
21h 31m 48s	+48°26' 00"	29'	1,062 Lyr	Easy

Description and Notes

Amazing symmetry is shown in M39, which is a large, loose cluster in Cygnus, appearing as a giant "V" shape. A total of 30 stars from magnitude 6.8 are associated with this cluster and many of these can be resolved in a small telescope of 4". I observed this object with my 8" and noted a nice double star in the center and the whole group shone intensely blue. M39 can be found 3° North of rho Cygni.

• M2

Other Name	Type	Const	Mag	Class
NGC 7089	Globular	Aquarius	6.47	II
RA	**Dec**	**Size**	**Distance**	**Rating**
21h 33m 29.3s	–00° 49' 23"	16'	37,000 Lyr	Medium

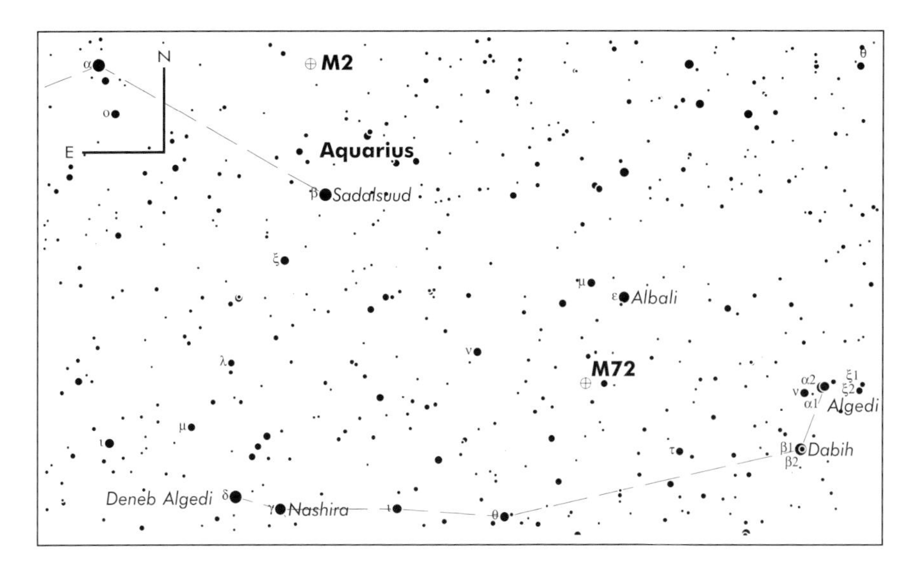

Figure 15.82. Finder Chart for Aquarius © Software Brisque.

Description and Notes

Aquarius has few prominent stars, but M2 can be tracked down by first locating the Magnitude 2.4 star epsilon Pegasi. Dropping down about 11° you will reach alpha Aquarii, then move 10° East to beta Aquarii, and M2 is located approximately 5° North West from this star. This cluster is fairly tight, compact in size and concentrated at the core. It sits within a fairly barren field, but with an 8" telescope I managed to resolve the cluster at 133×. If you spot M2, also try for the nearby globular M15 in Pegasus.

• IC 1434

Other Name	**Type**	**Const**	**Mag**	**Class**
–	Open	Lacerta	9.0	II 1 p
RA	**Dec**	**Size**	**Distance**	**Rating**
22h 10m 30s	+52°50' 00"	6'	–	Hard

Description and Notes

Lacerta contains a number of both bright and faint open clusters, and IC 1434 is a fairly dim example, situated 2° North West of the magnitude 4.4 star beta Lacertae. A medium or large telescope, and high power will resolve many of the

Figure 15.83. M2 © Tony O'Sullivan.

40 stars from magnitude 12, but few cluster members are visible using smaller instruments.

• NGC 7243

Other Name	Type	Const	Mag	Class
Caldwell 16	Open	Lacerta	6.4	IV 2 p
RA	**Dec**	**Size**	**Distance**	**Rating**
22h 15m 08s	+49°53' 54"	29'	2,634 Lyr	Easy

Description and Notes

Also in Lacerta is NGC 7243, the brightest open cluster in this constellation, at magnitude 6.4, that can be detected in binoculars under dark skies. Its component stars are mainly magnitude 8 and 9, and with a 4" telescope I resolved about 30 stars, 2.5° South West of beta Lacertae. A large instrument, 10" or more will increase this star count considerably to at least 50–60 members.

• NGC 7245

Other Name	Type	Const	Mag	Class
OCL 225	Open	Lacerta	9.2	II 1 p
RA	**Dec**	**Size**	**Distance**	**Rating**
22h 15m 11s	+54°20' 36"	5'	6,865 Lyr	Hard

Figure 15.84. IC 1434 © Tony O'Sullivan.

Description and Notes

In a 4" telescope NGC 7245 is difficult to detect, showing only a faint mist of unresolved stars. With a large aperture of 10" the member stars are rich and show a diffused haze towards the clusters middle. The brightest stars are around magnitude 13, so averted vision will improve detection. This cluster is positioned nearly 4° South of zeta Cephei.

• NGC 7510

Other Name	Type	Const	Mag	Class
OCL 256	Open	Cepheus	7.9	II 2 m n
RA	**Dec**	**Size**	**Distance**	**Rating**
23h 11m 03s	+60°34' 12"	6"	6,764 Lyr	Medium

Description and Notes

NGC 7510 is a small cluster in Cepheus where the bright stars forms an interesting curved "V" shape. The cluster can be observed in a 4" telescope, although using my 8" Newtonian I resolved 13 stars and detected nebulosity surrounding the cluster, though apparently this is illusory. Situated 7° North West of beta Cassiopeiae, NGC 7510 is one of my favorite small clusters, that I stumbled across whilst observing bright clusters in Cassiopeia.

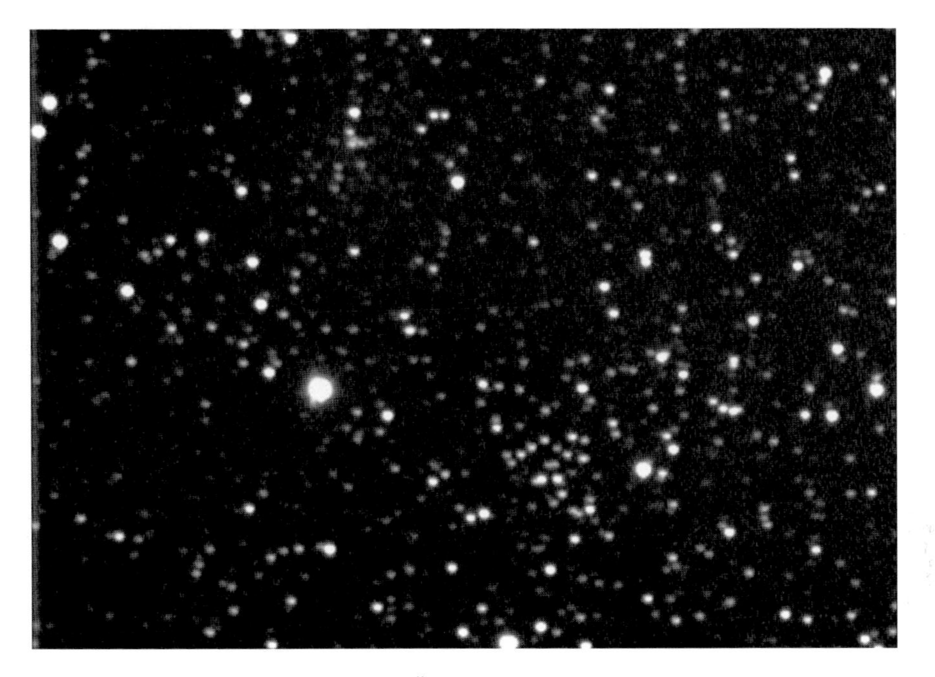

Figure 15.85. NGC 7245 © Tony O'Sullivan.

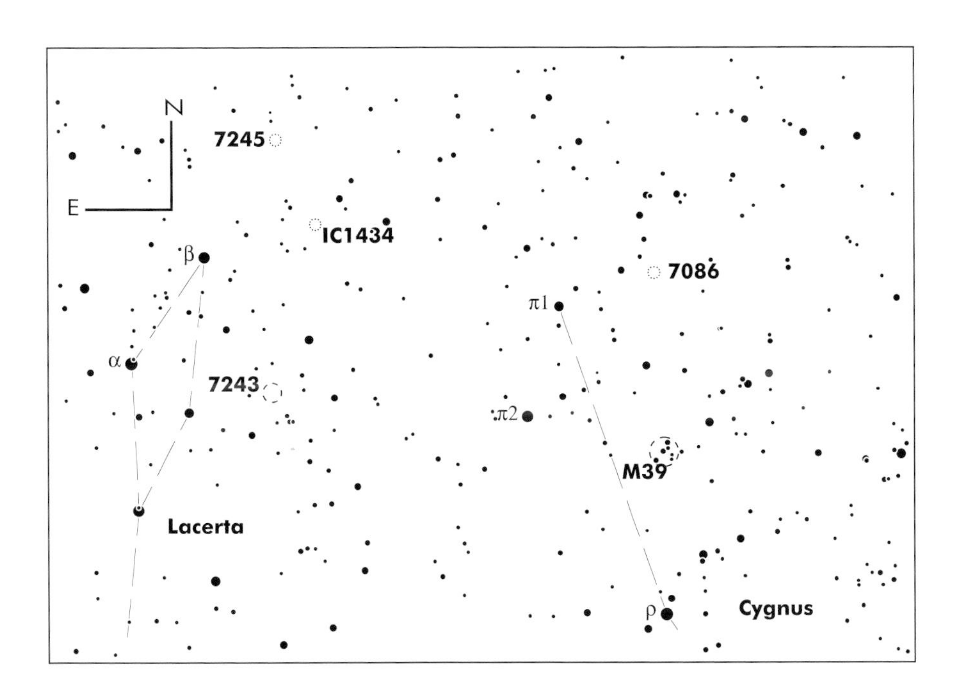

Figure 15.86. Finder Chart for Lacerta © Software Bisque.

• M52

Other Name	Type	Const	Mag	Class
NGC 7654	Open	Cassiopeia	6.9	I 2 r
RA	**Dec**	**Size**	**Distance**	**Rating**
23h 24m 48s	+61°35' 36"	15'	4,632 Lyr	Medium

Description and Notes

Cassiopeia contains many open clusters, but M52 is particularly stellar rich and dense, containing over 200 star members. There are 4 or 5 bright central stars, and in my 8" f6, I noted that the group appeared to have nebulosity around it, but this not the case. It is simply "star glow." At low power, M52 has several arcs of stars that appear to radiate from its center, and lies 6° North West of beta Cassiopeiae.

• King 21

Other Name	Type	Const	Mag	Class
–	Open	Cassiopeia	9.8	III 3 m
RA	**Dec**	**Size**	**Distance**	**Rating**
23h 49m 54s	+62°43' 00"	4'	6,855 Lyr	Difficult

Figure 15.87. M52 © Tony O'Sullivan.

Figure 15.88. King 21 © Tony O'Sullivan.

Description and Notes

King 21 is one of the more obscure open clusters in Cassiopeia and contains about 15 to 20 stars, located about 4° North West of beta Cassiopeiae. A 4" telescope shows very little, although 10" aperture or more will reveal most of the stellar members and some field stars, though they will still be faint.

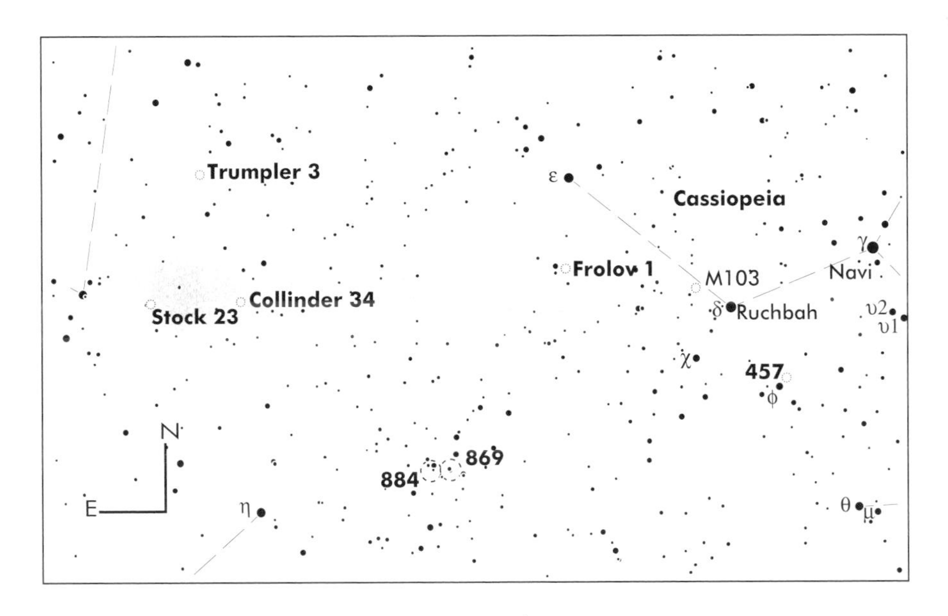

Figure 15.89. Finder Chart 1 for Cassiopeia © Software Bisque.

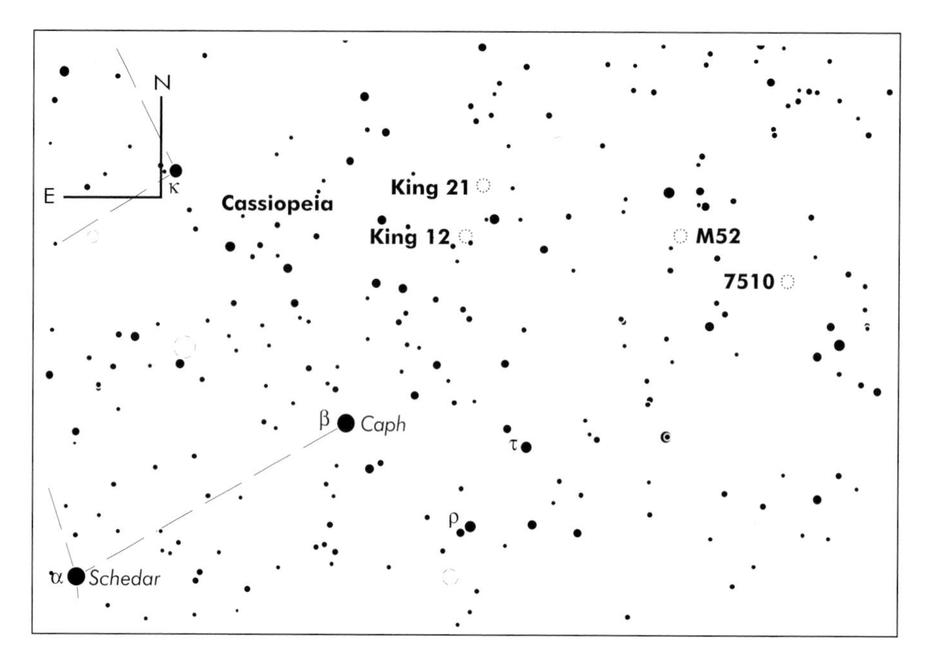

Figure 15.90. Finder Chart 2 for Cassiopeia © Software Bisque.

• King 12

Other Name	Type	Const	Mag	Class
–	Open	Cassiopeia	10	I2 p
RA	**Dec**	**Size**	**Distance**	**Rating**
23h 53m 00s	+61°58' 00"	3'	7,752 Lyr	Hard

Description and Notes

King 12 is another faint cluster that lies within the confines of Cassiopeia, although in a small telescope such as a 4" you may resolve a handful of very faint stars at high power. In reality, a larger telescope puts on a better show but do not expect a staggering view. This cluster lies about 3.5° North West of beta Cassiopeiae.

Figure 15.91. King 12 © Tony O'Sullivan.

• Frolov 1

Other Name	Type	Const	Mag	Class
–	Open	Cassiopeia	9.2	–
RA	**Dec**	**Size**	**Distance**	**Rating**
23h 57m 24s	+61°38' 00"	3'	–	Hard

Description and Notes

Frolov 1 is a tiny, curious little group of 4 or 5 stars visible in a 4" telescope, although a larger instrument will display several more members and reveal some of the fainter background stars. Located just under 3° North West of beta Cassiopeiae, the cluster itself is sparse and the brightest member only magnitude 10.6, so the overall appearance is subtle.

Figure 15.92. Frolov 1 © Tony O'Sullivan.

Chapter 16

Catalogues and Cluster Data

The following catalogues and databases were used throughout this book, to obtain positional data and information on cluster size, distance, and magnitudes, as well as other related material.

Star Cluster Catalogues

These catalogues contain the latest object lists and data for all the known and suspected open and globular clusters within the Milky Way.

Catalogue of Parameters for Milky Way Globulars
William E. Harris, McMaster University
Harris, W. E. 1996, AJ, 112, 1487
http://physun.physics.mcmaster.ca/Globular.html
http://www.seds.org/messier/xtra/supp/mw_gc.html

New catalogue of optically visible open clusters and candidates
Wilton S. Dias, Universidade de Sao Paulo
Dias W.S., Alessi B.S., Moitinho A., Lepine J.R.D.
2002, A&A, 389, 871D
http://www.astro.iag.usp.br/~wilton/

Observational Data for Galactic Globular Clusters
Brian Skiff
Webb Society Quarterly Journal for January 1995, no. 99
http://www.ngcic.org/default.htm

The nonexistent star clusters of the RNGC
Brent Archinal
The Webb Society, Monograph No. 1
http://www.webbsociety.freeserve.co.uk/

General Catalogues

The catalogues listed here contain various different types of deep-sky objects, but also include many good examples of open and globular clusters. The Messier and Caldwell objects have been extensively covered in many publications and several of them are listed in Chapter 15. For a complete listing of the NGC and IC catalogues, refer to the NGC/IC Project website listed under "Star Cluster Data."

Messier Catalogue (M)	Messier, first published 1794, 110 Objects
New General Catalogue (NGC)	Dreyer, first published 1888, 7840 Objects
Index Catalogue (IC 1 and 2)	Dreyer, first published 1895, 5386 Objects
Caldwell Catalogue (C)	Moore, first published 1995, 110 Objects
Herschel 400	Herschel, 400 selected objects from the Herschel catalogues.

Specific Catalogues

Below is a list of both contemporary and modern cluster catalogues that are often based on a specific study or research program, and commonly named after the principal investigator involved. Where applicable, abbreviations are shown, which are often quoted in astronomy software and observing lists. Some of the clusters in these catalogues are listed in other publications under different names, such as the NGC. These lists are subject to change as new clusters are discovered or old objects are demoted into "non-cluster" status.

- **Open Clusters**

	Alessi
A	Antalova
Av-Hunter	Aveni-Hunter
Bar	Barkhatova
Bas	Basel
Be	Berkeley
Bi	Biurkan
	Blanco
Bo	Bochum
Cr	Collinder
	Chupina
Cz	Czernik
Do	Dolidze
DoDz	Dolidze/Dzimselejsvili
Fr	Frolov
Fein	Feinstein
	Graff
	Grasdalen
Haf	Hafner
H	Harvard
	Hogg
	Ivanov

Isk	Iskudarian
	Juchert
	King
	Kronberger
	Latysev
	Lynga
	Loden
Mrk	Markarian
	Mayor
Mel	Melotte
Pal	Palomar
Pi	Pismis
Ro	Roslund
Ru	Ruprecht
Steph	Stephenson
St	Stock
	Sher
	Teutsch
Tom	Tombaugh
Tr	Trümpler
Up	Upgren
	Waterloo
Westr	Westerlund
VdBerg	van den Bergh-Hagen

- **Globular Clusters**

	Arp
AM	Arp-Madore
BH	van den Bergh-Hagen
Dun	Dunlop
	Djorg
H	William Herschel
HP	Haute Provence
Lac	Lacaille
Pal	Palomar
Ter	Terzan
Ton	Tonantzintla
UKS	United Kingdom Schmidt

Star Cluster Data

Much of the data, statistics and other information were obtained from the following sources, which contain a wealth of material for amateur astronomers. These organizations produce and maintain extensive databases that are useful for checking and cross referencing star cluster observations.

NGC/IC Project

Core Team Members: Harold Corwin, Steve Gottlieb, Bob Erdmann, Malcolm Thomson and Wolfgang Steinicke.

Completely revised and updated NGC and IC catalogues are available, including observational data on many clusters.

http://www.ngcic.org

Saguaro Astronomy Club (SAC)

SAC Deep Sky Database 7.2

Database Team: Steve Coe, Thad Robosson, and A.J. Crayon.

A text format data file than can be imported into any database program, and contains invaluable information on over 10,000 deep-sky objects.

http://www.saguaroastro.org/

WEBDA–Star cluster database

This is a website devoted to stars in open and globular clusters.

obswww.unige.ch/webda/

SEDS–Students for the Exploration and Development of Space

SEDS is an independent, student-based organization that promotes the exploration and development of space.

http://www.seds.org/

SIMBAD Astronomical Database

Contains a wealth of statistical information on star clusters.

CDS - Centre de Données astronomiques de Strasbourg.

Operated at CDS, Strasbourg, France

http://simbad.u-strasbg.fr/

Other Data and References

The following publications and organizations were also used to research source and reference information.

NASA -Astrophysics Data System (ADS)

http://adswww.harvard.edu/

American Astronomical Society

http://www.aas.org/

ESO–European Southern Observatory

http://www.eso.org/

The Astrophysical Journal

http://www.journals.uchicago.edu/ApJ/

Astronomy and Astrophysics

http://www.edpsciences.org/aa

Please note that due to the evolutionary structure of the Internet, some of the web addresses listed may be subject to change.

Postscript

During our journey on the subject of star clusters, we have encountered a dramatic range of objects that continue to fascinate and amaze us. It is a testament to our complex makeup that we are able to understand and comprehend these celestial delights in the first place, and yet it is no coincidence that if they did not exist, neither would we. For this reason alone it could be argued that we are somewhat duty bound to study, observe and appreciate what the Universe has to offer.

Whilst writing this book, the experience highlighted several key issues and confirmed some of my initial beliefs. First, the topic of star clusters is so great that with current knowledge and theories it could easily fill hundreds of volumes, and although I have only just scratched the surface, I hope I have covered the most pertinent points and done the subject some justice. Second, I now know that writing a book is a daunting and challenging prospect, but is ultimately an enjoyable and fulfilling experience. Finally, I still firmly believe that star clusters are the most beautiful and rewarding objects in the night sky, and can provide the observer with a lifetimes worth of realistic targets.

My travels have now come to an end, but if the reader has gained an insight into these objects, and is inspired enough to go out and actually observe them, then my mission has not been in vain.

Index